Lecture Notes in Computer Science 6236

Commenced Publication in 1973
Founding and Former Series Editors:
Gerhard Goos, Juris Hartmanis, and Jan van Leeuwen

Burkhard Stiller   Tobias Hoßfeld
George D. Stamoulis (Eds.)

Incentives, Overlays, and Economic Traffic Control

Third International Workshop, ETM 2010
Amsterdam, The Netherlands, September 6, 2010
Proceedings

 Springer

Volume Editors

Burkhard Stiller
University of Zürich, Communication Systems Group
Department of Informatics
Binzmühlestrasse 14, 8050 Zürich, Switzerland
E-mail: stiller@ifi.uzh.ch

Tobias Hoßfeld
Universität Würzburg, Lehrstuhl für Informatik III
Am Hubland, 97074 Würzburg, Germany
E-mail: hossfeld@informatik.uni-wuerzburg.de

George D. Stamoulis
Athens University of Economics and Business, Department of Informatics
76 Patision Street, 10434 Athens, Greece
E-mail: gstamoul@aueb.gr

Library of Congress Control Number: 2010933119

CR Subject Classification (1998): C.2, H.4, D.2, H.3, D.4, C.2.4

LNCS Sublibrary: SL 5 – Computer Communication Networks and Telecommunications

ISSN   0302-9743
ISBN-10   3-642-15484-0 Springer Berlin Heidelberg New York
ISBN-13   978-3-642-15484-3 Springer Berlin Heidelberg New York

springer.com

© Springer-Verlag Berlin Heidelberg 2010
Printed in Germany

Typesetting: Camera-ready by author, data conversion by Scientific Publishing Services, Chennai, India
Printed on acid-free paper   06/3180

Preface

Economic perspectives in network management have recently attracted a high level of attention. The Third Workshop on Economic Traffic Management (ETM 2010) was the continuation of two successful events that were held at the University of Zürich, Switzerland in 2008 and 2009. The main objective of ETM 2010 was to offer scientists, researchers, and operators the opportunity to present innovative research on ETM mechanisms, to discuss new related ideas and directions, and to strengthen the cooperation in the field of economics–technology interplay. Being co-located with the International Teletraffic Congress (ITC22), ETM 2010 brought together a new and fast-growing scientific community.

The concept of ETM has emerged due to the fact that a multitude of different self-interested players are simultaneously active in the Internet. While such players may either compete or complement each other in the value chain for service providers, each of them has his own incentives and interests. To enable a win–win situation for all players involved (basically end users, Internet Service Providers (ISP), telecommunication operators, and service providers), new incentive-based approaches have been recently developed, tested, and even commercially deployed, which fall under the domain termed Economic Traffic Management (ETM). ETM mechanisms aim at improving efficiency within the network, e.g., by reducing costs, while also improving Quality-of-Experience (QoE) for end users or applications. In view of the increase of overlay traffic, driven amongst others by overlay and peer-to-peer (P2P) applications, more traditional and global optimization approaches, e.g., route optimizations or network management, tend to be superseded by ETM solutions. Such solutions take into account interactions among various players and employ mechanisms that can lead the system into a viable equilibrium. That is, while each player reacts according to his own interests to the mechanism, e.g., in terms of traffic inserted, information disclosed, or Quality-of-Service (QoS) levels selected, the design of the latter is such that the system is led to a mutually beneficial situation, without having to assume any further coordination.

ETM is particularly applicable to cases involving thousands or even millions of individual users injecting traffic into networks of multiple interacting network service providers, possibly acting on different tiers and pursuing different incentives. Due to the decentralization of these players and due to the commercialization of service offerings, a scalable and economically driven approach offers a wider range of interesting alternatives for optimization, traffic management, and network management, and taking care of respective legal views. Finally, besides these advantages, ETM also serves the increasing importance of socio-economic studies in the Future Internet, since its ultimate goal is the improvement of QoE for end users, yet in a way that is economically sustainable for providers, too.

The increased interest in these topics is confirmed by the fact that this year 21 papers were submitted to the ETM workshop. The authors of the papers come from 17 countries, with 82% from Europe, 12% from Asia/Pacific, and 6% from the USA. Out of them six papers were selected as full papers, following a thorough peer-review process. The acceptance rate was, thus, approximately at 29%, reflecting the very competitive nature of the workshop and the high quality of the work presented. In addition, ETM 2010 decided to accept four short papers on emerging ideas to stimulate fruitful discussions during the workshop.

The major focus of the papers submitted, according to the authors' selection of topics, is seen in the following topics, namely: (1) Economic management of traffic and its related economic implications, (2) ETM mechanisms and technologies, (3) ETM application scenarios, such as that of peer-to-peer applications, overlay networks, or virtual networks, (4) application-layer traffic optimization (ALTO), and (5) economically efficient bandwidth allocation. This was also reflected by this year's program of papers, organized in a session on "P2P and Overlay Management" and "Evaluations and Estimations" followed by the Short Paper session. The program was complemented by the keynote presentation on "Socio-economic Challenges for the Internet of the Future: The Case of Congestion Control" given by Costas Courcoubetis, Athens University of Economics and Business.

Many people contributed a great amount of their time to organize ETM 2010. Therfore, special thanks are addressed to the Technical Program Committee and all additional reviewers. Further thanks go to Rob van der Mei and Hans van den Berg, contacts to the ITC 22 hosting conference, to Cristian Morariu, Universität Zürich, for supporting the ETM 2010 website dynamics, to Evelyne Berger, Universität Zürich, for handling all registrations and pre-workshop matters, and to Thomas Zinner, University of Würzburg, for his support during the workshop.

Finally, the editors would like to address their thanks to Springer namely, Anna Kramer, for a smooth cooperation on finalizing these proceedings. Additionally, many thanks go to the support of the European FP7 STREP project "Simple Economic Management Approaches of Overlay Traffic in Heterogeneous Internet Topologies (SmoothIT)," No. 216259 and the FP7 NoE on "Anticipating the Network of the Future — From Theory to Design (Euro-nf)," No. 216366.

September 2010

Burkhard Stiller
Tobias Hoßfeld
George D. Stamoulis

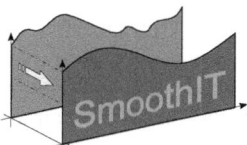

Organization

General Chair

Burkhard Stiller   University of Zürich, Switzerland

Program TPC Co-chairs

Tobias Hoßfeld   University of Würzburg, Germany
George D. Stamoulis   Athens University of Economics and Business, Greece

Publicity Chair

Zoran Despotovic   DOCOMO Communications Labs Europe, Germany

Publications Chair

Piotr Chołda   AGH University of Science and Technology, Poland

Web Master

Andrei Vancea   University of Zürich, Switzerland

Technical Program Committee

Eitan Altman   INRIA, France
Dominique Barth   University of Versailles, France
Torsten Braun   Universität Bern, Switzerland
Maria Angeles Callejo   Telefonica Investigacion y Desarrollo, Spain
Costas Courcoubetis   Athens University of Economics and Business, Greece
György Dan   KTH Stockholm, Sweden
Philip Eardley   Britisch Telecom, UK
Markus Fiedler   BTH Karlskrona, Sweden
David Hausheer   UC Berkeley, USA and
                                    University of Zurich, Switzerland
Nikolaos Laoutaris   Telefonica Investigacion y Desarrollo, Spain
Nicolas Le Sauze   Alcatel-Lucent Bell Labs, France
Antonio Liotta   Eindhoven University of Technology, The Netherlands
Marco Mellia   Politecnico di Torino, Italy

Konstantin Pussep	Technische Universität Darmstadt, Germany
Peter Racz	University of Zürich, Switzerland
Peter Reichl	Telecommunications Research Center Vienna, Austria
Sergios Soursos	Intracom Telecom R&D, Greece
Spiros Spirou	Intracom Telecom R&D, Greece
Dirk Staehle	University of Würzburg, Germany
Rafal Stankiewicz	AGH University of Science and Technology, Poland
Bruno Tuffin	INRIA, France
Kurt Tutschku	University of Vienna, Austria

Reviewers

A number of detailed and highly constructive reviews for papers submitted to ETM 2010 were made by all of our reviewers, comprising the Technical Program Committee (TPC) members as stated above and additionally George Exarchakos, Eindhoven University of Technology, Christian Groß, Technische Universität Darmstadt, Frank Lehrieder, University of Würzburg, Vlado Menkovski, Eindhoven University of Technology, Simon Oechsner, University of Würzburg, and Piotr Wydrych, AGH University of Science and Technology.

Therefore, it is of great pleasure to the Technical Program Co-chairs to thank all these reviewers for their important and valuable work.

Table of Contents

Socio-economic Challenges for the Internet of the Future: The Case of Congestion Control

Costas Courcoubetis

Athens University of Economics and Business Athens, Greece
courcou@aueb.gr

Abstract. The Internet is founded on a very simple premise: sharing! Shared communications links are more efficient than dedicated connections that lie idle much of the time. Hence the rules we use for sharing are extremely vital for the healthy operation of the Internet ecosystem and directly affect the value of the network to its users. It becomes a great paradigm of merging the disciplines of computer science and economics, and presents a great number of challenges to the Internet research community.

In this talk we will discuss a number of questions like: what is wrong with todays Internet sharing technologies? Are these consistent with economics? More specifically, is TCP (Transmission Control Protocol) sensible from an economic point of view? Which network sharing technologies justify end-toend from an economics perspective? What is required to make P2P (peer-topeer) a blessing instead of a curse? Are there bad applications or just inefficient combinations of sharing technologies and pricing schemes?

This keynote will survey some of these systems management issues and the emerging approaches to solve them.

B. Stiller, T. Hoßfeld, and G.D. Stamoulis (Eds.): ETM 2010, LNCS 6236, p. 1, 2010.

An Incentive-Based Approach to Traffic Management for Peer-to-Peer Overlays

Konstantin Pussep[1], Sergey Kuleshov[2], Christian Groß[1], and Sergios Soursos[3]

[1] Multimedia Communications Lab, TU Darmstadt, Germany
{pussep,chrgross}@kom.tu-darmstadt.de
[2] PrimeTel R&D, Cyprus
sergeyk@prime-tel.com
[3] Intracom Telecom R&D, Greece
souse@intracom.com

Abstract. Peer-to-Peer overlays are responsible for a large amount of consumer traffic, including the costly inter-domain traffic. Managing this traffic requires to consider the interests of the involved parties (users, overlay providers, and network providers), since many traditional approaches benefit only single parties. In this work we propose a mechanism where an Internet provider offers additional *free* resources to selected users that act in a most network-friendly way and are able to bias the overlay traffic for higher localization. By the means of simulations, we show that a proper cooperation with overlay providers can result in a mutual benefit. For typical interconnection agreements, this also applies if only a single Internet provider adopts the proposed mechanism.

1 Introduction

Management of Peer-to-Peer (P2P) traffic has traditionally been a challenging task as it requires meeting the diverse requirements of all involved parties: users, overlay providers, and Internet Service Providers (ISPs). The main goal of the users is the increased performance and quality of experience. Overlay/content providers strive for decreased load on their servers by increasing content availability in the overlay and, insofar, their goals match those of the users, since higher availability allows for the better performance for the users. Finally, ISPs look to minimize their costs, particularly those incurred from inter-domain traffic. With the increasing popularity of high-resolution video content [1] the relevance of mechanisms matching the interests of all three parties is even more crucial.

Traditional techniques to manage Internet traffic focus on prioritization of certain protocols (e.g., HTTP over P2P traffic) and traffic shaping [2] to meet the requirements of ISPs and some users. However, they do not take into account the needs to the growing number of P2P users and their impact on the network neutrality is highly debatable [3]. Other approaches, such as proxy caching, face the limitation to support the vast amount of protocols and potential legality issues. Finally, some approaches try to minimize inter-domain traffic by localizing the overlay traffic inside the ISP's domain. However, the locality itself might lead

B. Stiller, T. Hoßfeld, and G.D. Stamoulis (Eds.): ETM 2010, LNCS 6236, pp. 2–13, 2010.

neither to increased overlay performance nor content availability, thus, the goals of the overlay users and provider are not necessarily met [4].

In this work an approach to indirect Traffic Management (TM) is presented which attempts to meet the requirements of all three parties by giving users a clear and measurable *incentive* to cooperate both with the ISP, in terms of promoting locality, and with overlay providers, in terms of increasing bandwidth availability. The proposed method boosts the overlay performance in the local domain by increasing the upload bandwidth of selected local peers. These peers are referred to as *Highly Active Peers* (HAP) and are promoted by means of the Next Generation Networking (NGN) or similar features of the ISP's network management system. The mechanism also increases the download bandwidth of HAPs as an additional incentive to promote locality and upload more data to other local peers.

We show that, though all peers can benefit from improved overlay performance, best results are achieved if the overlay implements locality-awareness techniques, which would allow other peers to discover local HAPs easier and benefit from their increased capacities. We also pay special attention to the *early adopter* scenario when only one ISP applies the proposed technique.

This paper is structured as follows: The related work is discussed in Section 2, while Section 3 provides the overview of the proposed approach including the incentive mechanism, monitoring, and promotion technique. In the subsequent Section 4 the application scenario is described. Finally, Section 5 presents the evaluation results, followed by the conclusion in Section 6.

2   Related Work

Traffic shaping is one possible technique to reduce the costly inter-domain traffic incurred through P2P applications. A prominent example is the case of Comcast that throttled BitTorrent traffic [8]. However, such measures typically result in unsatisfied users and overlay providers. Additionally, their impact on network neutrality is highly debatable.

Another option are network caches [9] tailored to specific P2P applications such as those offered by Oversi[1]. Such caches typically rely on Deep Packet Inspection or similar techniques [6] in order to detect P2P traffic and fail to handle encrypted P2P traffic. Furthermore, the network cache owner might face problems, if caching illegal content.

A different approach is to enhance P2P applications with knowledge about the underlying infrastructure, thus, making them *locality-aware*. Different approaches were proposed, mostly applicable to the BitTorrent protocol [5,10,11]. Some techniques rely on collaboration with the network provider, offering locality hints [12,7] while others apply external sources of locality information [13]. As pointed out by recent work, such mechanisms not necessarily benefit all players [4]. Therefore, additional mechanisms that complement locality-awareness at the ISPs' side become interesting.

---

[1] http://www.oversi.com/

3   Approach Overview

The performance of P2P overlays for content distribution (such as file sharing and streaming) relies strongly on two factors: availability of the content and the upload bandwidth of participating peers. While the overlay performance only considers these factors globally, the Internet providers should attempt to improve them in the local domain. Then and only then can they successfully promote locality without hurting any of the involved stakeholders but leading to the so-called Triple-Win situation when peers, overlay, and ISPs benefit. Attempts to promote locality without first improving content availability and locally available bandwidth might lead to overall performance deterioration of the overlay.

The proposed mechanism suggests to increase locally available upload bandwidth by increasing upload bandwidth for the selected subset of overlay users who are most likely to increase the overlay performance and to promote locality for the benefit of the ISP. Consequently, the mechanism focuses on promoting the best overlay users to Highly Active Peers (HAPs) by significantly increasing their bandwidth for sufficient time intervals.

The proposed mechanism can be broken down to following steps:

1. The ISP monitors the performance of its overlay users and collects their usage statistics by the means described in Section 3.2.
2. Based on these statistics aggregated at regular intervals ISP calculates ranking metric and decides which peers to offer a potential to act as HAPs (see Section 3.1).
3. ISP uses NGN capabilities to promote selected peers by increasing their upload bandwidth (see Section 3.3 for details). At the same time HAPs might receive other benefits, such as increased download bandwidth, which can be also modified via NGN capabilities.

It is important to note that the promotion, and any associated benefits, stays active until the next evaluation phase takes place. Then the peers that already have been HAPs in the previous phase can either remain HAPs if they have behaved as expected, or be demoted to normal peers otherwise.

3.1   HAP Selection Algorithm

The goal of the algorithm is to identify *heavy users*, being able to act as locality-promoting HAPs and bias the overlay traffic for more locality.

3.1.1   Ranking Metric

The metric for selecting peers to be promoted to HAPs has to consider relative contribution of the peers to the overlay network as a whole as well as to the local domain, thus addressing the interests of Internet and overlay providers, which are not always aligned. As peers can have rather diverse configurations, all evaluations have to be done relative to the maximum possible contribution in any given period of time, thus keeping the metric fair to all potential HAPs. Based on the above requirements the following parameters describe peer's contribution to overlay in global and local domains (calculated over the measurement interval t with a typical value of 24 hours):

- $U_{total}(t)$ is the total upload traffic of the peer. This metric represents total contribution of the peer to the overlay and is important to consider for the benefit of overlay since peers who already contributed a lot, could potentially contribute even more.
- $U_{local}(t)$ is the total upload traffic of the peer within the local domain. This parameter allows to prefer peers that probably apply locality-aware peer selection mechanisms or that are popular inside the domain, being more quickly discovered by other local peers.
- $S(t)$ is the seeding ratio, i.e., the ratio of peer's upload and download volume. It allows to select peers that are altruistic in the sense that they stay online to provide content and not only to consume it (this allows us to avoid the measurement of actual online time of peers, that is practically rather difficult). $S_{max}(t)$ is the highest value over all peers during time slot t.
- $B(t)$ is the upload bandwidth of the peer during time slot t.
- d is the duration of the measurement interval and depends on the minimum possible promotion interval and monitoring equipment capabilities (one day or less appears feasible).

A proper combination of these parameters can only be identified based on the extensive experimental and exploitation results. As a starting point the naive approach of taking the weighted sum of *relative* contribution appears sufficient:

$$R(t) = p_t \frac{U_{total}(t)}{d \cdot B(t)} + p_l \frac{U_{local}(t)}{d \cdot B(t)} + p_s \frac{S(t)}{S_{max}(t)} \tag{1}$$

Note that to keep the metric fair for all customers and agnostic to actual customer's bandwidth the upload volume is divided by its upper bound $d \cdot B(t)$ that would have been observed if a peer continuously uploaded data at full speed. Furthermore, it allows for fair comparison between HAPs and normal peers.

The exact values of weight parameters p_t, p_l, p_s depend on the requirements and context of the specific ISP deploying the mechanism. Generally, it is advisable to keep values of all weights relatively close in order to balance gain for both the Internet and overlay providers.

Another requirement to the HAP selection metric is that the history of the peers' behavior has to be taken into account as peers with a known history of high contribution are more likely to maintain the trend. On the other hand historical data should have limited effect over time in order to keep the entrance barrier quite low for new peers. A feasible approach is taken which suggests using the Simple Moving Average over duration of history period H which yields the following metric:

$$R = \frac{\sum_{t=t_{now}-H}^{t_{now}} R(t)}{H} \tag{2}$$

3.1.2 Promotion of HAPs

Based on the computed rating values, the ISP performs the promotion of peers/customers based on the presented ranking metric. The technical details and considerations of this operation are discussed in Section 3.3. Promotion is only active for the duration of time interval t, since during the next interval a new set of HAPs is selected, while peers not eligible to be HAPs again are reverted back to their normal access profile.

There are at least two different ways to determine the desired subset of peers to be promoted to HAPs:

1. The ISP decides on the additional upload and download bandwidth U_{all} and D_{all} that should be allocated to all HAPs and the per peer upload increase U_{peer} and D_{peer}. Then the total number of supported HAPs is:

$$N = min(U_{all}/U_{peer}, D_{all}/D_{peer})$$

2. Alternatively, the ISP defines a rating threshold R_{th} and promotes all peers with higher ratings.

We opt for the first solution, since it allows the ISP to limit the potential traffic increase in its network, though, a combination of both approaches can be used that promotes up to N best peers whose ratings exceed the threshold R_{th}. This choice is based on considerations of financial feasibility as every extra HAP not only leads to potential savings (due to increased locality of P2P traffic) but also brings some associated costs.

3.2 Collection of Behavioral Statistics

The basic parameters required to calculate the ranking metric as described in Section 3.1.1 can be collected in two ways: through the monitoring functionality of the ISP's network equipment such as NetFlow[2], or from the overlay application itself. Both approaches have their advantages and disadvantages some of which depend on the specific equipment in use by the ISP.

The first option makes the HAP mechanism application agnostic as it only requires deployment on the ISP side. In order to achieve best results, the ISP should be able to classify the exchanged data according to the protocol type: P2P, HTTP, etc. This can be done by the means of deep packet inspection or other techniques [6], which might be difficult for some ISPs. Otherwise, the collected statistics will contain the data about the total traffic of the given ISP's customer, resulting in a certain deviation. This can be tolerated to some extent, since most non-P2P applications are rather making use of download than upload bandwidth.

Collecting behavior information from the client requires additional cooperation between the overlay provider and the ISP. Unfortunately, this information cannot be fully relied upon as it can be potentially tampered and mislead the ISP into promoting wrong users. This can be partially addressed by cross checking the provided information, which, however, complicates the procedure.

---

[2] http://www.cisco.com/en/US/products/ps6601/
products_ios_protocol_group_home.html

3.3 Means of Changing the User Access Profiles

The incentive-based traffic management technique described in this work takes advantage of the capabilities provided by NGN compatible equipment, which, among other features, allows on-the-fly automated updates of the customer access profiles[3]. In our case the totally accessible upload and download bandwidth of certain users in increased. The same results, however, can be achieved using other equipment. One realistic example would be the case when customer access bandwidth is not limited by the DSLAM but is throttled by the use of a customized Linux-based traffic shaper, which can be reconfigured dynamically to promote certain customers to HAPs.

4 Application Scenario

Our approach targets, in the first place, content distribution overlays that make extensive use of the available peer bandwidth, that means especially file sharing and streaming application. Besides the potentially extensive usage of inter-domain links, such applications state significant requirements on the available upload bandwidth in order to satisfy users' requirements, such as fast downloads in case of file sharing and low playback delays for streaming. While the traffic share of pure file sharing overlays is currently decreasing, the streaming applications are increasingly gaining ground [1]. Here, we are especially interested in the peer-assisted Video-on-Demand (VoD) applications because they exhibit the properties of shifting the distribution costs from content providers to users and ISPs. While the users might not suffer significantly due to flat-rate based payment schemes, an ISP has to deal with the increased load on its links.

Therefore, a proper traffic management mechanism should be able to mitigate the negative effect of such cost shifts in the most efficient manner. This can be compared with the situation of Content Delivery Networks (such as Akamai[4]) that make special agreements with content providers and place their servers at major interconnection links. In our approach the ISP and the overlay provider might benefit in a similar manner by combining the locality-awareness in the overlay with the ISP-granted bandwidth.

A generalized peer-assisted application works as follows: The content is initially placed on overlay provider's servers and uploaded to requesting users. In case that the requested content was already consumed by some users, these replica owners are online and have free upload resources, new requests are served by those peers in the first place. Only the missing upload bandwidth is contributed through the servers. The amount of required bandwidth is dictated either by the desired download speed (for file sharing) or video bitrate (for streaming applications). This way, the content provider offloads it servers as much as possible and reduces its costs, while the users receive the desired Quality-of-Service.

---

[3] http://www.itu.int/ITUT/studygroups/com13/ngn2004/working_definition.html
[4] www.akamai.com

5   Evaluation

In order to evaluate the proposed mechanism, we apply event-driven simulations based on the application scenario described in Section 4. The simulation model captures all three involved players: users consuming and exchanging videos, overlay provider's servers offering the initial content, and network providers. The overlay application is able to run both in locality-*unaware* and -*aware* modes by applying different peer selection policies (similar to [5] and [7]).

The goal of our experiments is to understand whether the proposed approach can result in a "win-win" situation both for the overlay and ISPs. We address this by assessing the impact of HAP promotion through ISPs on the overlay performance and underlay costs (Section 5.2). For comparison, we consider the impact of locality-aware peer selection (Section 5.3) without HAP promotion. Furthermore, we evaluate the case where both options (HAP promotion with locality-aware peer selection) are combined (Section 5.4). This appears to be the most promising combination, since locality-awareness typically benefits the ISPs in the first place [5], while the HAP by itself increases the overlay resources.

A special case is when only one ISP decides to support HAP promotion. In such a case the impact on its inter-domain traffic must be considered, since having a clear benefit for such an *early adopter* increases the chance of the mechanism to be accepted by ISPs (see Section 5.5).

5.1   Experimental Setup and Metrics

The overlay comprises 10,000 users with 2,000 videos being consumed over the duration of four weeks. The videos are 50 minutes long and encoded with 610 kbps (considered as standard resolution for up-to-date video-on-demand services). We apply the Pareto distribution to user requests, resulting in 20% of users generating 80% of streaming requests. The users choose videos to watch according to a Zipf distribution (with Zipf parameter $\alpha = 0.85$) and stay online for 30 minutes after the playback finishes. Furthermore, the users store downloaded videos in their local caches limited to the size of two GB (enough to cache four videos) and apply a first-in-first-out cache replacement policy.

The network topology is modeled by six ISP domains: one containing only the content provider's servers and others containing the normal peers. The upload bandwidth of peers is limited to 305, 610, or 1220 kbps per scenario (meaning 0.5, 1 or 2 times the playback rate of the video, respectively). Since the typical download bandwidth of DSL users is at least 1.5 Mbps, we limit the download rate only to the video playback rate, e.g., peers never consume the stream at a higher rate than the playback rate. We don't limit the upload bandwidth of the servers, since we are interested in their load under various traffic management configurations. Note that we don't include the confidence intervals, since they were negligible (around 1%) for all experiments.

We capture the following metrics to understand the impact of the mechanisms under study: (1) *data uploaded by servers* to capture the costs for the overlay provider, and (2) *inbound and outbound inter-domain traffic* to capture

ISPs' costs. We also consider the *intra-domain traffic*, that does not leave ISPs' domain, as the best case regarding the interconnection costs.

Note that we don't address the user experience directly, since in a peer-assisted scenario the servers catch up with missing resources. Instead, we assume that the content provide transfers the distribution costs (or cost savings) to the users by adjusting the content prices. A feasible scenario is also one where the content provider rewards users depending on their overlay contribution.

5.2   Plain HAP

In this experiment we consider the case when all ISPs hosting normal peers apply the HAP mechanism. Starting with three different initial peer upload capacities, the upload bandwidth of discovered *heavy users* is increased by a factor of 1.5, 2, and 3, resulting in up to 200% more upload bandwidth in the overlay.

Figure 1(a) shows that, surprisingly, the increased upload bandwidth does not improve the interconnection costs of the ISPs significantly (only 8% decrease in the case of low initial bandwidth and 3x bandwidth increase per HAP). The reason is that the additional bandwidth is equally distributed in the overlay and, therefore, the probability that the additional bandwidth of *local peers* is being used, is quite low. On the other hand, Figure 1(b) shows a significant improvement in the server load, with savings of up to 90% for the content provider. We reason that the plain HAP mechanism does not necessarily results in savings for the ISPs, though being beneficial for the content provider.

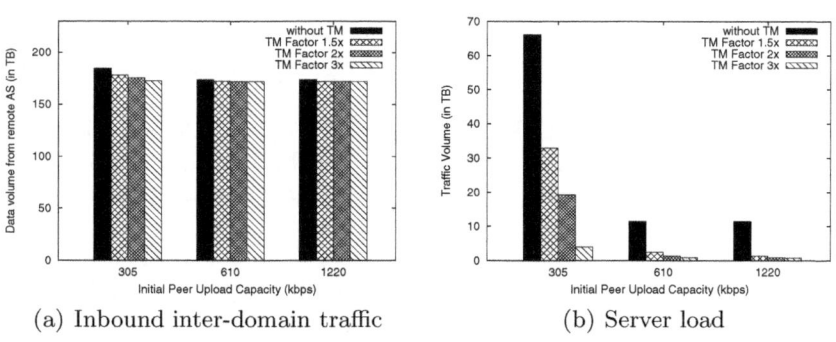

(a) Inbound inter-domain traffic   (b) Server load

Fig. 1. Plain HAP promotion applied by all ISPs

5.3   Locality-Aware Peer Selection

Following the insights of the previous experiment, we consider the case of an ISP-friendly overlay application, where the peers always try to download video segments from local peers, if possible. As shown in Figure 2, an ISP-friendly overlay design allows to reduce both inbound and outbound traffic of ISPs without having an ISP-side traffic management in place. This effect is stronger if the upload bandwidth of peers is not scarce (i.e. equal or larger than the video

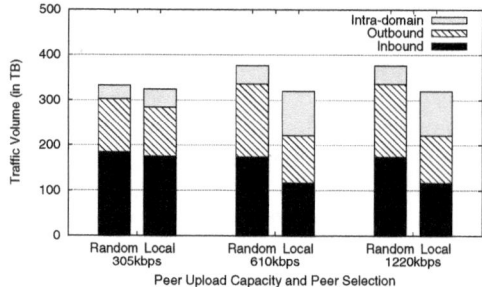

Fig. 2. Impact of locality-aware peer selection (without HAP promotion)

bitrate). Unfortunately, the locality-aware peer selection does not result in a benefit for the overlay as such, since the load on the servers cannot be reduced (we assume that there are no bottlenecks in the inter-domain links). Therefore, the overlay provider has no clear incentive to implement the locality-aware peer selection in the application.

5.4 HAP with Locality-Aware Peer Selection

The results of the previous experiments suggest the combination of the HAP mechanism (applied by ISPs) and the locality-aware peer selection (applied by the overlay) being promising to achieve a win-win situation. Hereby, both parties have to contribute their part to the mutual benefit.

Figure 3(a) shows the potential savings of the inbound traffic. The observed reduction ranges from 30 to 50% depending on the bandwidth increase for HAPs. On the other hand, the server load is reduced by up to 90% (see Figure 3(b)), similar to the case of the HAP approach and locality-unaware overlay (cf. Figure 1(b)). Note that this reduction of both the inter-domain traffic and server load is achieved by promoting only 20% of peers to HAPs.

To conclude the experiment, an ISP benefits from the HAP mechanism, if the boosted peers use an ISP-friendly overlay application. This can be done, for example, by preferring HAP candidates uploading more data inside of the ISP's domain than to remote ISPs (by setting the weight $p_l \gg p_t$ and p_s).

5.5 Early Adopters

In this experiment we analyze the situation of an early adopter, i.e., a single ISP that applies HAP mechanism while the other ISPs don't apply it yet. Similar to the previous experiments, we compare the impact of HAP promotion for locality-unaware and locality-aware overlays, expecting the second type of application to result in better ISP-friendliness. However, it is not obvious whether the early adopter can benefit from the HAP promotion or might experience even higher inter-domain traffic.

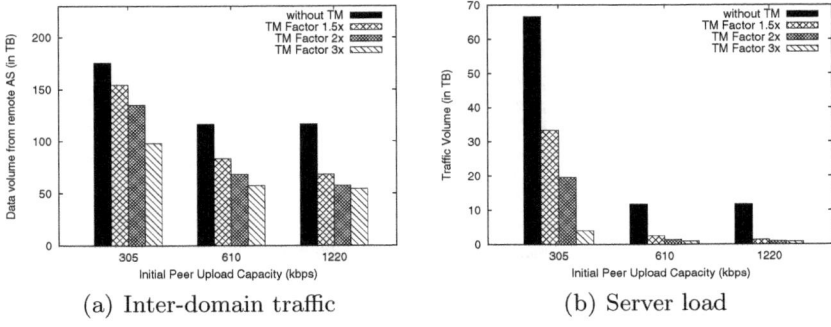

(a) Inter-domain traffic   (b) Server load

Fig. 3. HAP applied by all ISPs combined with locality-aware peer selection

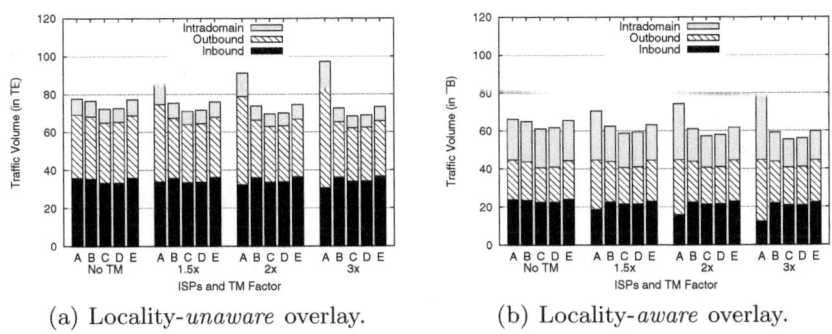

(a) Locality-*unaware* overlay.   (b) Locality-*aware* overlay.

Fig. 4. Traffic management applied only by the *single ISPs (A)* for both locality-unaware and -aware overlays. Initial upload bandwidth of peers is fixed to 610 kbps.

Figure 4 shows the impact on inter-domain and intra-domain traffic of all five ISPs where only ISP A applies HAP promotion. Here the peers' upload bandwidth is the same as the video bitrate. We show the impact of different values of the HAP TM factor (1 = no traffic management, 1.5, 2, or 3 times the original upload bandwidth) on the inter- and intra-domain traffic of single ISPs.

In all cases we observe a reduction in *inbound* traffic and increase in *outbound* traffic for ISP A. For other ISPs the HAP adaptation by ISP A results in a moderate decrease of the outbound traffic, while the inbound traffic stays almost the same. We interpret the situation of the early adopter ISP depending on its interconnection agreements with neighbor ISPs as follows:

- If A pays only for the *inbound* traffic, the HAP mechanism can result in a significant cost reduction. Here the overlay-incurred inbound traffic is reduced by up to 50% (see Figure 4(b)). We consider this as a typical situation of tier-3 ISPs.
- If A pays for the sum of both inbound and outbound traffic, the HAP mechanism will result in increased interconnection costs since the total inter-domain traffic increases for the early adopter (ISP A).

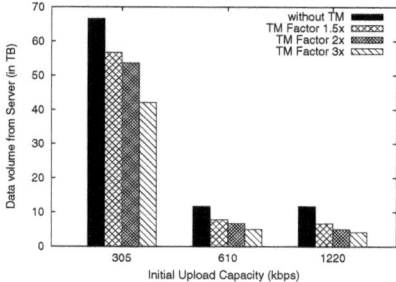

Fig. 5. Server load if only ISP *A* applies HAP

– If A *is paid* by smaller ISPs for A's outbound traffic, the early adopter might even increase its interconnection income. This especially applies, if ISP *A* is a tier-2 ISP, while others are tier-3 ISPs.
– If A has peering agreements with neighbor ISPs, the payments themselves are not affected, though the long term impact of changed traffic profiles might result in changed interconnection agreements.

In summary, the early adopter of the HAP mechanism wins in most cases, except the unlikely case when his costs are dominated by the outbound traffic. We further observe that the ISP *A*'s situation is best if: the overlay applies locality-aware peer selection and the upload bandwidth increase per peer is sufficiently high (> 50% decrease for TM factor of 3 as shown in Figure 4(b)).

Additionally, Figure 5 shows the impact of HAP promotion applied only by ISP *A* on the server load. As expected, the load reduction is lower compared to all ISPs applying traffic management (up to 50% compared to 90% in Figures 1(b) and 3(b)). Since only the users in ISP *A*'s domain receive increased upload bandwidth, the *total* overlay bandwidth increases only moderately and results in lower server load savings compared to the case when all ISPs apply HAP. Nevertheless, the overlay provider is able to save 30-50% of its costs, resulting in a win-win situation for the overlay and ISP *A*.

6   Conclusion

This paper has presented a promising mechanism to bias overlay traffic for more locality to the benefit of P2P overlays and ISPs. In the proposed mechanism an ISP increases the upload bandwidth of selected users to improve the overlay performance and to significantly reduce ISP's interconnection costs. The selection of suitable peers takes into account their global and local upload behavior.

The considered application scenario is a peer-assisted Video-on-Demand system, where an overlay provider tries to minimize the load on its servers while providing high streaming experience to the users. The overlay provider has the freedom to apply locality-aware peer selection policy that would result in reduced interconnection costs for the ISPs and increase the chance of its users being promoted. We further show that the mechanism can be effective, even if applied by a single ISP, especially when the inbound traffic dominates the interconnection costs.

Our future work will concentrate on the dynamics of HAP selection including the impact of parameter weights. Additionally, we will address the financial feasibility and socio-economical impact of proposed incentives on ISPs and users.

Acknowledgments

This work has been performed in the framework of the EU ICT Project SmoothIT (FP7-2007-ICT-216259). The authors would like to thank all SmoothIT partners for useful discussions on the subject of the paper.

References

1. Cisco Visual Networking Index: Forecast and Methodology, 2008–2013, White Paper (2009), http://www.ciscosystemsverified.biz/en/US/netsol/ns827/networking_solutions_sub_solution.html
2. Nair, S., Novak, D.: A traffic shaping model for optimizing network operations. European Journal of Operational Research 180(3), 1358–1380 (2007)
3. Peha, J.: The Benefits and Risks of Mandating Network Neutrality, and the Quest for a Balanced Policy. International Journal of Communication (2007)
4. Piatek, M., Madhyastha, H., John, J., Krishnamurthy, A., Anderson, T.: Pitfalls for ISP-friendly P2P design. In: Hotnets (2009)
5. Oechsner, S., Lehrieder, F., Hoßfeld, T., Metzger, F., Pussep, K., Staehle, D.: Pushing the Performance of Biased Neighbor Selection through Biased Unchoking. In: 9th International Conference on Peer-to-Peer Computing (2009)
6. Karagiannis, T., Papagiannaki, K., Faloutsos, M.: BLINC: Multilevel Traffic Classification in the Dark. In: SIGCOMM (2005)
7. Aggarwal, V., Feldmann, A., Scheideler, C.: Can ISPS and P2P Users Cooperate for Improved Performance? SIGCOMM Comput. Commun. Rev. 37, 29–40 (2007)
8. Comcast Throttles BitTorrent Traffic, Seeding Impossible (2007), http://torrentfreak.com/comcast-throttles-bittorrent-traffic-seeding-impossible/
9. Saleh, O., Hefeeda, M.: Modeling and Caching of Peer-to-Peer Traffic. In: IEEE International Conference on Network Protocols (2006)
10. Bindal, R., Cao, P., Chan, W., Medved, J., Suwala, G., Bates, T., Zhang, A.: Improving Traffic Locality in BitTorrent via Biased Neighbor Selection. In: 30th International Conference on Distributed Computing Systems (ICDCS 2006) (2006)
11. Wang, J., Huang, C., Li, J.: On ISP-friendly Rate Allocation for Peer-sssisted VoD. In: International Conference on Multimedia. ACM, New York (2008)
12. Xie, H., Krishnamurthy, A., Silberschatz, A., Yang, Y.R.: P4P: Explicit Communications for Cooperative Control Between P2P and Network Providers. In: SIGCOMM (2008)
13. Choffnes, D.R., Bustamante, F.E.: Taming the Torrent. In: SIGCOMM (2008)

Quantifying Operational Cost-Savings through ALTO-Guidance for P2P Live Streaming

Jan Seedorf[1], Saverio Niccolini[1], Martin Stiemerling[1],
Ettore Ferranti[2], and Rolf Winter[1]

[1] NEC Laboratories Europe, Kurfuerstenanlage 36, 69115 Heidelberg, Germany
{seedorf,niccolini,stiemerling,winter}@nw.neclab.eu
[2] ABB Corporate Research, Segelhofstrasse 1K, 5405 Baden-Dattwil, Switzerland
ettore.ferranti@ch.abb.com

Abstract. Application Layer Traffic Optimization (ALTO) is a means for operators to guide the resource provider selection of distributed applications. By localizing traffic with ALTO, Internet Service Providers (ISPs) aim to reduce the amount of costly traffic between Autonomous Systems (ASes) on the Internet. In this paper, we study the potential cost-savings for operators through ALTO-guidance for a specific type of P2P application: P2P Live Streaming. We use datasets that model the Internet's AS-level routing topology with high accuracy and which estimate the business relationships between connected ASes on the Internet. Based on this data, we investigate different ALTO strategies and quantify the number of costly AS-hops traversed.

Our results show that indeed transmission costs can be reduced significantly for P2P Live Streaming with ALTO. However, for this particularly delay-sensitive type of application, ISPs have to be careful not to *over-localize* traffic: if peers connect to too many peers which are in the same AS but have low upload capacity, chunk loss increases considerably (resulting in poor video quality). In addition, we demonstrate that if ISPs use an ALTO strategy which recommends peers solely based on the transmission costs from the ISP's perspective, neither the individual ISPs nor the overall system can substantially decrease transport costs.

1 Introduction

Application Layer Traffic Optimization (ALTO) [13] is currently being investigated as a promising means to improve resource provider selection for distributed applications. By providing certain information, operators can help applications in choosing where to download a desired resource (which is available at multiple locations in the network) from. Examples of the kinds of information which can be useful to convey to applications via ALTO are operators' policies, geographical location (e.g. network proximity), or the transmission costs associated with sending/receiving a certain amount of data [12]. In short, ALTO - as envisioned by the IETF - is a dedicated service, operated by a network operator or *Internet Service Provider (ISP)*, which can provide useful network layer information

B. Stiller, T. Hoßfeld, and G.D. Stamoulis (Eds.): ETM 2010, LNCS 6236, pp. 14–26, 2010.

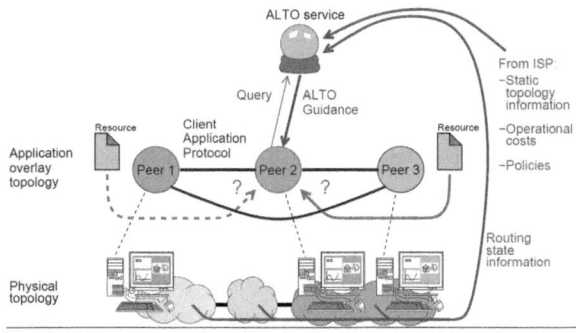

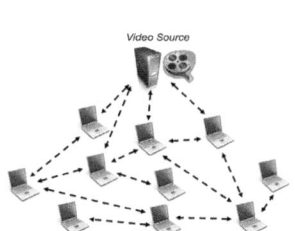

Fig. 1. Improving Peer Selection with ALTO   **Fig. 2.** P2P Live Streaming

to application layer clients about resource providers. One class of applications which can particularly benefit from ALTO are P2P applications. For instance, receiving information about the underlying network layer topology enables P2P applications to take closeness on the physical topology into account when choosing their neighbors. Figure 1 [13] exemplifies ALTO-guidance for P2P applications: A peer can request information about several candidate peers (which offer a desired resource, e.g., a chunk of a file) from an ALTO-service. Through ALTO, an ISP can recommend certain candidate peers, depending on information the peer could not derive itself. In the example, it is likely that the ALTO service would recommend to download from **peer 3** as this peer is in the same network layer domain as **peer 2**. ISPs can thus use an ALTO-service to influence the peer selection process of P2P applications to localize traffic. One core goal of such traffic localization is to save costs for the ISP, i.e., the transmission costs of data items sent between different *Autonomous Systems (ASes)*.

In this paper, we study the potential cost-savings for operators through ALTO-guidance for a specific type of P2P application: P2P Live Streaming. Such services have recently appeared on the Internet; popular examples include PPLive, Sopcast, or TVAnts. P2P Live Streaming is different from file-sharing in in that it requires low delay and high, constant bandwidth in order to ensure that chunks arrive within a certain threshold (i.e. the *playout delay* of the stream buffer at the peers). To account for these requirements, P2P Live Streaming applications employ dedicated chunk scheduling algorithms [1]. As a consequence of these properties, traffic localization for P2P Live Streaming is more challenging compared to File-Sharing applications because optimized peer selection may interfere with the live streaming requirement of receiving chunks with low delay. If by using ALTO, peers connect to too many local peers (to reduce costly cross ISP traffic), these peers may not be optimal from the application's point of view in delivering chunks with low delay (e.g. because these peers have low upload bandwidth which correlates with high delay). Our goal is to study the effects of ALTO with respect to such *over-localization* of traffic.

The main contributions of this paper are: 1) We investigate different peer selection algorithms and ALTO strategies to study the trade off between traffic localization, cost savings for the operator, and video stream user experience, 2) We quantify the actual cost savings through ALTO-guidance for P2P Live Streaming applications using sophisticated datasets of the Internet's AS-level routing topology which also estimate the business relationships between ASes on the Internet, and 3) We demonstrate that if all ISPs use ALTO to recommend peers solely based on the associated transmission costs for downloading a chunk (from the downloading peer's ISP's perspective), neither the individual ISPs nor the overall system can decrease transmission costs significantly. In summary, our results show that for P2P live streaming applications with reasonable video bitrates, operational costs can significantly be reduced with ALTO-guidance. However, for this particularly delay-sensitive application ISPs have to be cautious not to *over-localize* traffic, which results in too many lost chunks for the application. We analyse the trade off between cost-savings and the risk of over-localization for different ALTO strategies in detail. The rest of this paper is organised as follows. Section 2 describes the system model we use to analyse ALTO and P2P Live Streaming as well as our model of the Internet's AS-level routing topology. In section 3 we present and discuss the results of our experiments. Section 4 discusses related work and highlights our contribution in relation to existing work. Section 5 concludes the paper with a summary and an outlook on future work.

2   System Model

In a P2P Live Streaming system, a source node splits the video stream into chunks and distributes these chunks to initial peers in the overlay. Then chunks are exchanged among peers according to a chunk scheduling strategy (push/pull). Peers use a *peer selection algorithm* to find peers to connect to initially and to frequently refresh their neighbor list (i.e. the list of peers chunks are exchanged with). Peers select which exact chunks to send/receive to/from peers in their neighbor list based on a *chunk scheduling algorithm* (e.g. *Latest Blind Chunk*, *Latest Useful Chunk* [1]). To study the effect of ALTO-guidance for such a system and the effect of ALTO on inter-AS traffic, we use the model detailed in this section.

Overlay and Chunk Scheduling Model. The focus of our study is on peer selection and cost savings through ALTO, and not to investigate different chunk scheduling strategies. We therefore model a generic *mesh-pull P2P Live Streaming* system as depicted in figure 2: Periodically, peers query their neighbors, asking which chunks these can provide; peers then select where to download which chunk from and sends out download requests accordingly to the respective peers. More precisely, we have a set P, consisting of N peers. Each peer $p_i \in P$ has unlimited download bandwidth (to model that this is not the bottleneck for receiving chunks timely) and $v(p_i)$ upload bandwidth. Initially, each peer $p_i \in P$ has degree d (randomly chosen[1]) neighbors, which form its neighbor set $D(p_i)$. In addition, each

---

[1] e.g., retrieved from a tracker.

peer $p_i \in P$ has a request buffer $r_b(p_i)$ for received chunk requests, and a chunk buffer $c_b(p_i)$ of size C chunks it received from other peers. Chunks older than a certain *playout-delay* are discarded from $c_b(p_i)$, i.e. the chunk buffer represents a sliding window, and C = playout-delay. Frequently (every *pulling-interval*), each peer p_i asks all its d neighbors $\in D(p_i)$ which chunks they currently have in their chunk buffer and for the size of their request buffer. For each missing chunk c_x, p_i requests the chunk among all peers which can offer c_x from the peer $p_j \in D(p_i)$ with $min(size(r_b(p_j)))$. This download request is added to the request buffer of p_j, $r_b(p_j)$. Old requests (i.e., ones where p_i has found a peer with smaller request buffer to download from) are discarded by the pulling peer: The respective upload peer is notified and removes the request from its request buffer. Peers sequentially upload chunks according to their request buffer[2]. Each peer p_i downloads from peer $p_j \in D(p_i)$ which first offers to upload a needed chunk, independent of the chunk position in the stream.

Network Layer and Transport Costs Modeling. Based on the focus of our analysis and the sheer size of the Internet routing topology we use a model of the network topology at the AS-level with over 30,000 ASes [14]. The actual topology is based on a large set of observed Internet routing data including e.g. BGP routing tables and updates. From the data, business relationship were inferred which we use as input into our simulation. These business relationships can be either of a customer-provider nature (paid peering) or based on a mutual peering agreement (settlement-free). The model comes as a set of matrices that allow to infer AS paths through the Internet, distances and business relationships. These matrices have been carefully created to reflect the actual Internet topology with a high degree of accuracy.

Peer Selection Algorithms. We model a round-based traffic optimization: At the beginning of every epoch e_t (specified in number of chunks), each peer p_i evaluates its neighbors. Depending on its *link-refresh algorithm*, p_i keeps k peers and discards all other peers from $D(p_i)$. The missing $d - k$ peers are replaced by a) obtaining $r * d$ new random peers (where r is a parameter that determines how many new candidate peers are retrieved), b) sorting these *candidate peers* according to a *peer-sorting algorithm*, and adding the top $d - k$ peers from the sorted list to the neighbor set $D(p_i)$. Each peer is in charge of refreshing links, and in general we investigate the following link-refresh algorithms: 1) *Throughput-based Link Refreshing* - each peer p_i keeps a link to another peer $p_j \in D(p_i)$ only if a chunk has been downloaded from p_j in the previous epoch e_{t-1}, or 2) *Locality-based Link Refreshing* - each peer p_i keeps a link to another peer $p_j \in D(p_i)$ only if p_j is in the same Autonomous System (AS) as p_i. The discarded peers are either deleted or added to the list of (new random) candidate peers which are being sorted to fill up the missing $d - k$ links in $D(p_i)$. The latter strategy allows for

---

[2] Note that this is not a limitation because the upload bandwidth is still fully utilised. However, chunks get diffused in the overlay quicker with sequential uploads (compared to parallel uploads) because as soon as a complete chunk has been uploaded, this particular chunk can be further diffused by the peer which downloaded it. [4]

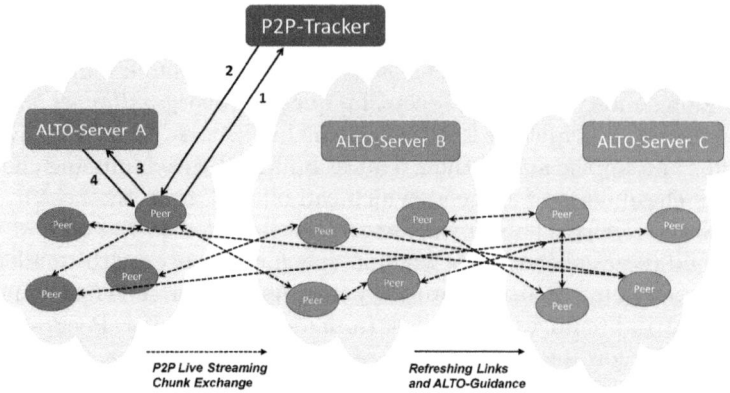

Fig. 3. ALTO-Guidance in a P2P Live Streaming System with a tracker

convergence of the sorting algorithm: Peers which were rated high in sorting but did not fulfill the link-refresh criterion are sorted again, together with all new candidate peers (i.e. $r*d+(d-k)$ peers are sorted). Peers with a high ranking are thus likely to prevail in $D(p_i)$ over several epochs. The sorting of new candidate peers is done in conjunction with an ALTO server and discussed in the following subsection.

ALTO-Guidance. As highlighted in the previous subsection, every epoch e_t peers obtain $r*d$ new candidate peers. Peers can retrieve new candidate peers either from a tracker or through gossiping protocols from other peers [8]. Our model is orthogonal to the source of new candidate peers, i.e. our results apply to ALTO-sorting regardless where the peers obtain new candidate peers from. Figure 3 displays ALTO sorting for the former case: Each peer contacts a centralized tracker (1) to retrieve new candidate peers (2). It sends the list of these candidate peers to the ALTO server of its ISP (3) which returns a sorted list of the candidate peer based on a peer sorting algorithm (4). We assume that peers use the recommendation from ALTO in peer selection [12]; cases where peers ignore ALTO guidance are out of scope of this paper.

We assume that peers know the available upload bandwidth of other peers and make this information available to the ALTO server (e.g. in step 3 in figure 3). Peers can either estimate upload bandwidth through measurements or peers can make this information available to other peers (e.g. publishing the upload bandwidth at the tracker or using a dedicated repository [4])[3]. Further, based on our Internet topology model (see section 2), we assume that the ALTO server has two kinds of information available to guide peers: the number of AS-hops between

---

[3] Note that the IETF ALTO working group [10] is currently discussing whether *provisioned upload bandwidth* can be information an ALTO server should provide. In this case, peers could use their ALTO server to obtain this information and then publish it, or the downloading peer would need to contact the uploading peer's ALTO server to retrieve this information.

two peers, or the (monetary) transmission costs for the peer's ISP associated with receiving a chunk from another peer. We only consider the downloading peer's ISP's transmission costs (i.e. the business relationship of the last inter-AS hop from the downloading peer's perspective) because this is the only information each ISP has available to guide peer selection in its network. ISPs normally are not aware of the business relationships details between two other ASes [14].

Thus, an ALTO-server can use the following information as a basis for sorting a set of candidate peers $Cand(p_i) = p_{j=1...r*d}$ for a downloading peer p_i: a) each candidate peer's upload bandwidth $v(p_j)$, b) the number of AS-hops between p_i and p_j, and c) the associated transmission costs (for the ISP of p_i) if p_i would download a chunk from p_j. We postpone the discussion and analysis of ALTO sorting based on the third type of information (i.e. the downloading ISP's transmission costs) to section 3. To combine the first two types of information (i.e. upload bandwidth and number of AS-hops), we analyse the following peer-sorting algorithms with input parameters α and β:

- **Disjoint-Bucket Sorting (DB)**: Candidate peers get sorted in *disjoint buckets* and $0 < \alpha < 1$, $0 < \beta < 1$, and $\alpha + \beta = 1$. Bucket A sorts all candidate peers based on the number of AS-hops between p_i and the candidate peer (giving lower values preference). Bucket B sorts all candidate peers solely based on the upload-bandwidth of the candidate peer (giving higher values preference). The sorting function returns $\alpha * (d - k)$ peers from bucket A and $\beta * (d - k)$ peers from bucket B. In other words, of the $(d - k)$ missing peers a fraction of α peers is returned solely based on AS-hop distance, and a fraction of β peers is returned solely based on upload bandwidth of the candidate peer.
- **Weighted-Sum Sorting (WS)**: Candidate peers get sorted based on a *normalized weighted sum* as follows[4], returning peers with lower values on top of the sorted list:

$$WS(p_j) = \alpha \times AShops_{norm}(p_i, p_j) + \beta \times v_{norm}(p_j) \qquad (1)$$

$$AShops_{norm}(p_i, p_j) = \frac{AShops(p_i, p_j)}{max_{AShops}(P)} \qquad (2)$$

$$v_{norm}(p_j) = \frac{\frac{1}{v(p_j)}}{max\left(\frac{1}{v(P)}\right)} \qquad (3)$$

The rationale behind the weighted sum is to account for a compromise between upload bandwidth and AS-hop distance when ranking candidate peers. Indeed, some peers may have a good (but not excellent) upload bandwidth as well as a fairly low (but not extremely low) AS-hop distance. Such peers may be preferable over peers with very high upload bandwidth but very long AS-hop distance (or

---

[4] For the weighted sum there need not be restrictions on α and β, but for simplicity we use again $0 < \alpha < 1$, $0 < \beta < 1$, and $\alpha + \beta = 1$.

AS-hops / Upl.-Bw.	1	2	3	4	5	6	7	8
5,000	0,045	0,070	0,095	0,120	0,145	0,170	0,195	0,220
1,000	0,127	0,152	0,177	0,202	0,227	0,252	0,277	0,302
0,384	0,292	0,317	0,342	0,367	0,392	0,417	0,442	0,467
0,128	0,825	0,850	0,875	0,900	0,925	0,950	0,975	1,000

Fig. 4. Weighted Sum Peer Selection ($\beta = 0.8, \alpha = 0.2$)

AS-hops / Upl.-Bw.	1	2	3	4	5	6	7	8
5,000	0,065	0,115	0,165	0,215	0,265	0,315	0,365	0,415
1,000	0,127	0,177	0,227	0,277	0,327	0,377	0,427	0,477
0,384	0,250	0,300	0,350	0,400	0,450	0,500	0,550	0,600
0,128	0,650	0,700	0,750	0,800	0,850	0,900	0,950	1,000

Fig. 5. Weighted Sum Peer Selection ($\beta = 0.6, \alpha = 0.4$)

vice versa), but such peers are unlikely to be returned as the top peers by the disjoint bucket sorting algorithm.

Using the reciprocal of the upload bandwidth ensures that peers with higher upload bandwidth get a lower value in equation 3 (and thus are preferred). By dividing through the maximum value in equation 2 and 3 the values get normalized to a scale from 0 to 1. Figure 4 displays the results of the weighted sum for $\beta = 0.8, \alpha = 0.2$, AS-hops 1-8, and 4 specific peer upload capacities; figure 5 displays the same results for $\beta = 0.6$ and $\alpha = 0.4$. It can be observed that for instance a candidate peer with upload capacity of $5Mbps$ and a distance of 4 AS-hops will get a better (i.e. lower) ALTO-ranking (0.120) than a candidate peer with upload capacity $1Mbps$ and a distance of 1 AS-hop (0.127) for $\beta = 0.8, \alpha = 0.2$. However, if $\beta = 0.6$ and $\alpha = 0.4$ the latter peer will get a better ranking (0.127) compared to the former peer (0.215). Hence, β and α can steer the compromise between upload-bandwidth or AS-hop distance in ALTO sorting.

We further refine the weighted-sum algorithm by strongly preferring intra-AS links (i.e. the case where AS-hops = 0) in sorting as follows:

$$WS(p_j) = sameAS(p_i, p_j) + \alpha \times AShops_{norm}(p_i, p_j) + \beta \times v_{norm}(p_j) \quad (4)$$

$$sameAS(p_i, p_j) = \begin{cases} 0, & \text{if } (p_i, p_j) \text{ are in the same AS,} \\ 1, & \text{if } (p_i, p_j) \text{ are not in the same AS,} \end{cases} \quad (5)$$

This enforces strong traffic localization as peers within the same AS always get a lower value in the sum than other peers, thus the highest ranking in the sorting.

3 Experiments and Simulation Results

This section presents the results of simulations we conducted using the system model of the previous section. We compare ALTO peer selection algorithms and analyse the effect on chunk loss and operational cost savings in different scenarios, aiming to find the optimal trade off.

Simulation Settings. We extended the P2P Live Streaming simulator *P2PTV-sim* [5] from the EU project *Napa-Wine* with a network layer model at the AS-level. We integrated the Internet topology matrices described in section 2 into the simulator so that for each chunk received by a peer, it is known how many inter-AS links the chunk traversed and how many of these links are of a customer-provider nature (i.e. *costly* AS links). Further, these datasets enable us to simulate ALTO sorting based on AS-hop distance and on transmission costs as outlined in section 2. Peers are allocated uniformly distributed among a randomly selected ASes (in all our simulations, we set $a = 100$). We set intra-AS latency $l_{intra} = 20ms$ and inter-AS latency $l_{inter} = 50ms$. The transmission time t for a particular chunk c_i being uploaded by a peer p_u and downloaded by a peer p_d is thus determined as follows in our simulations:

$$t(c_i, p_u) = \frac{size(c_i)}{v(p_u)} + (AShops(p_u, p_d) \times l_{inter}) + l_{intra}(p_u) + l_{intra}(p_d) \quad (6)$$

We simulate a pulling approach as described in section 2 and with parameters resembling a common P2P live streaming system such as PPLive [7]. In all simulations we use the following settings (similar to the measurements obtained in [8] [7]): Chunksize = 0.04 Mb, Videorate = 0.5 Mbps, C (PlayoutDelay) = $8s$, degree $d - 50$, pulling-interval = $50ms$, request buffer size = 30, epoch-size = 100 chunks, r = 2 (i.e. fetching twice as many candidate peers as the degree). In addition, every pulling-interval peers check if there is currently a chunk being downloaded (i.e. in transit for the receiving peer but not yet arrived) and if this is not the case peers refresh their links. This resembles link refreshing not only every epoch but also when peers start to get disconnected from the P2P swarm. We use a upload-bandwidth distribution as follows (based on the measurements used in [2]): $0.15 \times N$ peers with $v(p_i) = 5$ Mbps, $0.25 \times N$ peers with $v(p_i) = 1$ Mbps, $0.4 \times N$ peers with $v(p_i) = 0.384$ Mbps, and $0.2 \times N$ peers with $v(p_i) = 0.128$ Mbps. As a metric for transport costs, we report the total number of *costly* AS-links which were traversed for diffusing all chunks over the whole P2P network, normalized to 1000 chunks to compare different experiments. To measure traffic localization, we report the percentage of chunks which were exchanged on intra-AS links as well as the AS-level hop distribution. As a metric for user experience, we report the average chunk loss per peer (i.e. the number of chunks not received in time with respect to the sliding window of the chunk buffer). We only report the cost results for algorithms where the average chunk loss per peer was less than 10% of the total amount of chunks simulated. We regard a higher chunk loss as unnacceptable from a user experience perspective.

Comparing different ALTO Sorting Strategies. As a reference algorithm, we simulate simple throughput-based link refreshing (see section 2) combined with disjoint-bucket sorting with $\alpha = 0.0$, $\beta = 1.0$ to resemble a state-of-the-art P2P Live Streaming system without ALTO-guidance (we refer to this algorithm as *WA, Without ALTO*). Note that this is an optimistic reference algorithm because current P2P Live Streaming systems only in part select peers based on upload bandwidth [7]. Further, we simulate locality-based link refreshing in conjunction with disjoint-bucket sorting (DB) or weighted-sum sorting (WS) with various parameterizations (expressed as $[\beta, \alpha]$). Figure 6 shows the overall number of costly AS-links traversed (normalized to 1000 chunks) in the P2P

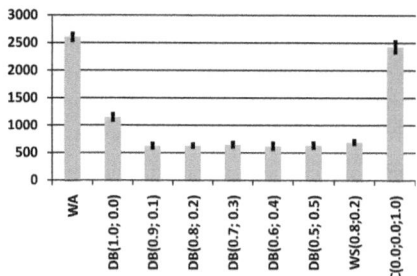

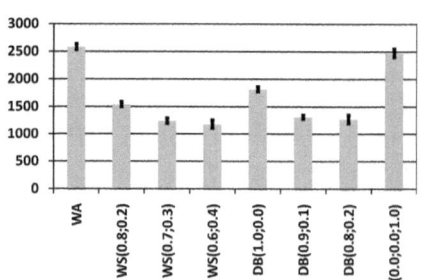

Fig. 6. Costly AS-hops traversed, 95 Confidence Interval (5000 peers, 2500 chunks)

Fig. 7. Costly AS-hops traversed, 95 Confidence Interval (1000 peers, 2500 chunks)

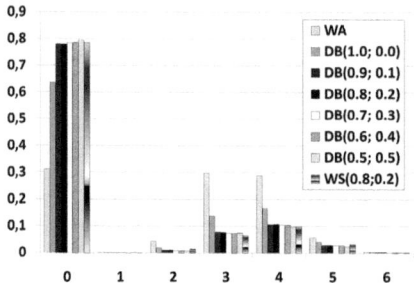

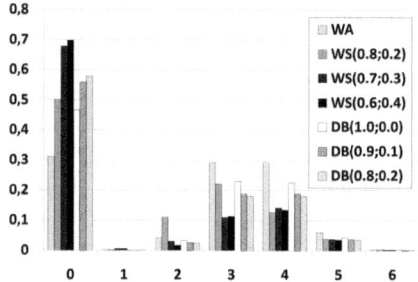

Fig. 8. Hop distribution for different algorithms (5000 peers, 2500 chunks)

Fig. 9. Hop distribution for different algorithms (1000 peers, 2500 chunks)

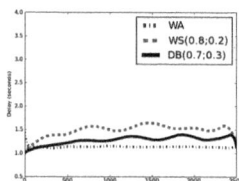

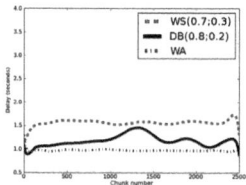

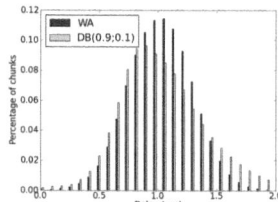

Fig. 10. Average chunk delay (5000 peers, 5000 chunks)

Fig. 11. Average chunk delay (1000 peers, 2500 chunks)

Fig. 12. Chunk delay distribution (5000 peers, 2500 chunks)

network for different strategies in a setting with 5000 peers. It can be observed that the ALTO strategies significantly reduce costs. However, there is not much difference among the different parameters. Figure 8 displays the corresponding hop distribution, showing that the algorithms achieve an intra-AS localization degree of nearly 80%. Also, it can be observed that the AS-hop distance (to peers not in the same AS) is mostly distributed between 3 and 5. Figure 7 shows the cost results similar to figure 6 but for 1000 peers. Since we keep the number of

Table 1. Average chunk loss per peer in number of chunks (5000 peers, 2500 chunks)

Algorithm	Chunk Loss
WA	0.10
WS(0.8;0.2)	0
DB(1.0;0.0)	120.97
DB(0.9;0.1)	0
DB(0.8;0.2)	0
DB(0.7;0.3)	0
DB(0.6;0.4)	0
DB(0.5;0.5)	0.15
C(0.0;0.0;1.0)	0

Table 2. Average chunk loss per peer in number of chunks (1000 peers, 2500 chunks)

Algorithm	Chunk Loss
WA	0
WS(0.8;0.2)	0
WS(0.7;0.3)	13.53
WS(0.6;0.4)	0.72
DB(1.0;0.0)	9.26
DB(0.9;0.1)	0
DB(0.8;0.2)	73.98
C(0.0;0.0;1.0)	0

Table 3. Average costs for ISPs of participating peers (2500 chunks)

Peers	Algorithm	Costs
5000	WA	111491.2
5000	C(0.0;0.0;1.0)	111153.0
5000	DB(0.8; 0.2)	25832.6
1000	WA	21901.4
1000	C(0.0;0.0;1.0)	22077.0
1000	WS(0.6;0.4)	10037.4

ASes and the degree constant ($a = 100$, $d = 50$), in this scenario peers cannot localize so much as compared to the scenario with 5000 peers because there are less peers in the same AS. It can be observed that in this scenario WS performs better than DB. Figure 9 provides the corresponding hop distribution, showing that WS(0.6;0.4) can achieve an intra-AS localization of 70%. Table 1 and 2 show the number of lost chunks for both scenarios. We observed that by increasing α, i.e. localization, further, chunk loss increases significantly. Figures 11 and 12 show that delay is not significantly affected by the ALTO-strategies we investigate (for strategies not displayed we obtained similar results).

We also conducted experiments with 5000 chunks to assure that the system has reached a steady state. We obtained similar per-1000-chunk costs as with 2500 simulated chunks, indicating that the system has already reached a steady state with 2500 chunks being simulated. Figure 10 confirms this, showing the average chunk delay for 5000 chunks for different strategies.

In summary, our results demonstrate that ALTO can significantly reduce costs while maintaining a low chunk delay. In addition, the results indicate that the optimal ALTO-strategy depends on the ratio between the degree and the number of peers in the same AS. If there are many peers in the same AS (runs with 5000 peers), there is the risk that peers connect almost exclusively to local peers. For instance, WS over-localizes if $\alpha \geq 0.3$, resulting in too many lost chunks. On the other hand, if there is less potential for intra-AS localization (runs with 1000 peers), WS can save slightly more costs than DB, and DB over-localizes for $\alpha \geq 0.3$. Furthermore, the detailed hop distributions reveal that - besides intra-AS localization - there seems to be not so much potential for traffic localization based on the number of AS-hops because the AS-hop distance on the Internet does not vary significantly enough to reduce delay nor costs based on AS-hop distance guidance.

Peer Selection based on inter-AS Business Relationships. In reality, some ISPs are likely to recommend candidate peers mainly based on the associated transmission cost from their perspective. To analyse this strategy, we simulated a disjoint bucket sorting similar to equation 1, but with a third bucket which sorts candidate peers solely based on the associated costs for the downloading peer's ISP. To study if such an approach is useful at all, we simulated the case where ISPs solely use this third bucket, which we call a C(0.0;0.0;1.0) strategy. It can be observed in figure 6 and figure 7, respectively, that in both scenarios this strategy performs only slightly better than without ALTO guidance at all (WA), and significantly worse than all other peer sorting algorithms. Additionally, table 3 shows not the overall system costs including intermediate AS-hops but the average costs for a peer's ISP, i.e. considering only costly AS-links traversed where one of the ASes has participating peers. Again, there are no cost savings and DB and WS perform much better. Looking into detailed simulation statistics revealed that the reason for this phenomenon is the underlying (realistic) cost model where a customer ISP pays for chunks on a customer-provider AS-link regardless of the traffic direction [14]. Hence, if a specific ISPs recommends to its peers to *download* form certain peers, it does not prevent that its peers *upload* via costly links. Ironically, such costly uploading may even be increased by ALTO guidance of the other downloading peer's ISP. Our results show that if all ISPs use this strategy, overall cost savings for the whole system as well as for the ISPs of participating peers are negligble.

4   Related Work

Guiding P2P applications with network layer information is not a completely new research topic. One of the first works to consider locality information provided by the ISP in order to improve P2P-traffic has been presented in [3]. A different approach which re-uses information provided by content delivery networks (CDNs) to guide peer selection for P2P-applications is proposed in [6]. There exist more studies, however, most approaches proposed in the literature have been evaluated for file-sharing applications and not for the particular challenges of live streaming. In particular, existing approaches do not consider a maximum delay in the range of seconds and the need to constantly receive a certain amount of chunks as required by P2P Live Streaming.

P2P Live Streaming applications have also received attention by researchers. Most research in this area is on measurements [7] and chunk scheduling [1], and not regarding optimal peer selection considering the underlying network layer topology. One exception is the work presented in [11] which proposes a decentralised peer selection algorithm based on the complete network layer transmission costs between peers. We consider it unrealistic for either peers or single ISPs to obtain the transmission costs for all AS-links a chunk traverses. In contrary, the focus of our study is on centralized ALTO-algorithms which use different types of information. In addition, our approach uses only the transmission costs for the peer's ISP in peer selection which is information a peer's ISP can provide to its ALTO-server.

One main contribution of our work is the AS-level modeling including the contractual relationships among the real Internet's ASes derived from sophisticated measurements. This enables to quantify the number of costly AS-links traversed on the real Internet for the proposed algorithms. Further, our work is the first to investigate how to combine different types of information in ALTO-guidance in order to prevent over-localization in P2P Live Streaming. To the best of our knowledge, our study is thus the first to analyse ALTO-algorithms for P2P Live Streaming applications and to quantify the potential cost savings based on a realistic Internet topology data set.

5 Conclusion

We investigated different peer selection strategies based on ALTO for P2P Live Streaming applications. Using an AS-level topology model of the real Internet, we provided quantitative estimations of the potential cost savings. These results demonstrate that a centralised ALTO approach can reduce overall transmission costs while keeping delay low, thus fulfilling streaming requirements. However, our analysis also reveals that ISPs have to be cautious not to over-localize traffic by recommending too many local but suboptimal peers. Further, our results indicate that the optimal peer selection strategy depends on the potential for localization in the system, i.e. the number of peers within the same AS. Finally, we showed that if ISPs use ALTO to recommend peers solely based on the associated transmission costs, there is no gain in cost reduction because ISPs' individual interests for traffic guidance may negate each other.

As future work, we consider to refine and analyse the proposed algorithms further, investigate more flexible ALTO-strategies which can adapt to over-localization dynamically, and to apply game theory to the case where ISPs recommend peers based on transmission costs in order to find a way to balance contradicting interests.

Acknowledgement. This work was partially supported by NAPA-WINE, a research project supported by the European Commission under its 7th Framework Program (contract no. 214412). The views and conclusions contained herein are those of the authors and should not be interpreted as necessarily representing the official policies or endorsements, either expressed or implied, of the NAPA-WINE project or the European Commission. The authors would like to thank Stella Spagna for help with extending the P2PTVsim simulator.

References

1. Abeni, L., Kiraly, C., Lo Cigno, R.: On the Optimal Scheduling of Streaming Applications in Unstructured Meshes. In: IFIP Networking 2009 (2009)
2. Bharambe, A.R., Herley, C., Padmanabhan, V.N.: Analyzing and Improving a Bit-Torrent Network's Performance Mechanisms. In: IEEE Infocom 2006 (2006)
3. Bindal, R., Cao, P., Chan, W., Medved, J., Suwala, G., Bates, T., Zhang, A.: Improving traffic locality in bittorrent via biased neighbor selection. In: ICDCS 2006, pp. 66–77 (July 2006)
4. Birke, R., Leonardi, E., Mellia, M., Bakay, A., Szemethy, T., Kiraly, C., Lo Cigno, R., Mathieu, F., Muscariello, L., Niccolini, S., Seedorf, J., Tropea, G.: Architecture of a Network-Aware P2P-TV Application: the NAPA-WINE Approach (under submission)

5. Couto da Silva, A.P., Leonardi, E., Mellia, M., Meo, M.: A Bandwidth-Aware Scheduling Strategy for P2P-TV Systems. In: IEEE P2P 2008 (2008)
6. Choffnes, D.R., Bustamante, F.E.: Taming the Torrent: A practical approach to reducing cross-isp traffic in peer-to-peer systems. ACM SIGCOMM Computer Communications Review (CCR) 38(4), 363–374 (2008)
7. Hei, X., Liang, C., Liang, J., Liu, Y., Ross, K.: Insights into PPLive: A Measurement Study of a LargeScale P2P IPTV System. In: IPTV Workshop, International World Wide Web Conference (May 2006)
8. Hei, X., Liu, Y., Ross, K.: IPTV over P2P streaming networks: the mesh-pull approach. IEEE Communications Magazine 46(2), 86–92 (2008)
9. Ng, T.S.E., Zhang, H.: Predicting internet network distance with coordinates-based approaches. In: IEEE Infocom, New York, vol. 1, pp. 170–179 (June 2002)
10. Peterson, J., Gurbani, V., Marocco, E., et al.: ALTO Working Group Charter, http://www.ietf.org/html.charters/
11. Picconi, F., Massoulie, L.: ISP-friend or foe? Making P2P live streaming ISP-aware. In: ICDCS 2009 (2009)
12. Seedorf, J., Burger, E.: Application-Layer Traffic Optimization (ALTO) Problem Statement. RFC 5693 (October 2009)
13. Seedorf, J., Kiesel, S., Stiemerling, M.: Traffic Localization for P2P-Applications: The ALTO Approach. In: IEEE P2P 2009 (September 2009)
14. Winter, R.: Modeling the Internet Routing Topology with a Known Degree of Accuracy - in less than 24h. In: ACM/IEEE PADS 2009 (2009)

Overlay Connection Usage in BitTorrent Swarms

Simon Oechsner, Frank Lehrieder, and Dirk Staehle

University of Würzburg, Institute of Computer Science, Germany
{oechsner,lehrieder,dstaehle}@informatik.uni-wuerzburg.de

Abstract. The amount of peer-to-peer (P2P) traffic and its inefficient utilization of network resources make it a high priority for traffic management. Due
to the distributed nature of P2P overlays, a promising approach for a management scheme is to change the client behavior and its utilization of overlay connections. However, from the viewpoint of a client, there are different categories
of overlay connections. In this paper, we discern between these different types of
overlay connections in BitTorrent, the currently most popular P2P file-sharing application. A simulation study of BitTorrent and a video-streaming derivate called
Tribler provides insights into the usage of these types of connections for data exchange. Thus, traffic management based on client behavior can be optimized by
efficiently targeting connections which carry the most traffic. We also show the
implications of these results for locality-awareness mechanisms such as Biased
Neighbor Selection and Biased Unchoking.

1 Introduction

While the share of traffic generated by peer-to-peer (P2P) file-sharing is decreasing,
its absolute amount still increases according to recent studies [CS09]. Moreover, video
streaming, which is the projected next top bandwidth consumer, is also partially being
conducted via P2P overlays. On the other hand, since P2P traffic is created by end hosts
and not by servers, it is much more difficult to control and to manage. Typically, overlays are underlay-agnostic, i.e., they do not consider the underlying network structure
in maintaining overlay connections. Therefore, they are comparably wasteful with network resources and therefore also with provider costs. As a result, management schemes
for this traffic are currently a major research topic.

The fact that the end user has to willingly participate in such a management scheme is
acknowledged since unilateral approaches from providers to throttle P2P traffic have led
to a decrease of customer satisfaction. Thus, these schemes have to provide an advantage to end users or at least must not lead to a disadvantage. This is formulated, e.g., in
the Economic Traffic Management (ETM) concept of SmoothIT [GRH+08, OSP+08].
Approaches following this guidelines are Biased Neighbor Selection (BNS) [BCC+06]
or Biased Unchoking [OLH+09]. Such mechanisms show that influencing decisions at
the clients can be an efficient way to manage P2P traffic, in contrast to detecting and
re-routing inefficient traffic flows in the core network.

While the most popular file-sharing overlays, such as BitTorrent, form random
meshes, not all overlay connections are the same from a specific client's point of view.
These different types of connections can be influenced differently by management

B. Stiller, T. Hoßfeld, and G.D. Stamoulis (Eds.): ETM 2010, LNCS 6236, pp. 27–38, 2010.
© Springer-Verlag Berlin Heidelberg 2010

schemes. Therefore, it is necessary to determine which of these connections carry more traffic, in which direction this traffic flows and which parts of the overlay generate the most traffic.

In this paper, we evaluate the connection usage in BitTorrent swarms, providing insights where traffic management at the clients should be most effective. This information serves as input to ETM discussions such as existing in the IETF working group for Application Layer Traffic Optimization (ALTO). The paper is structured as follows. In the next section, we provide background information on BitTorrent and a video streaming variant, namely Tribler. In the same section, we also review related work. Next, we describe the model for the simulative performance evaluation of these overlays. The results from these evaluations are given in Section 4. We conclude the paper by summarizing our findings in Section 5.

2   Background and Related Work

In this section, we first describe the mechanisms of interest for our evaluation, both of the BitTorrent and the Tribler protocol. These are the neighbor selection and unchoking mechanisms, which are both in the focus of current approaches to ETM. Then we review related work on the topic.

2.1   Key Mechanisms of BitTorrent

The BitTorrent protocol forms a mesh-based overlay and utilizes multi-source download to distribute content. For each shared file, one overlay is formed, a so-called *swarm*. To facilitate the multi-source download, a shared file is split into smaller pieces which are also called *chunks*. These chunks are in turn again separated into sub-pieces or *blocks*. A detailed description of BitTorrent can be found in [LUKM06] and [Bit]. In the following, we will focus on the description of the relevant mechanisms of BitTorrent that are modified for ETM.

Neighbor Set Management. Each peer has only a limited number of other peers it has direct contact with in the swarm. These neighbors know about each other's download progress, i.e., which chunks the other has already downloaded. This enables a peer A to signal its interest in downloading chunks to a neighbor B holding chunks that the local peer still is missing. We say that peer A is *interested* in peer B.

A peer joining a swarm typically initializes its neighbor set by contacting a *tracker*, i.e., an index server with global information about the peer population of a swarm. A standard tracker responds to queries with a random subset of all peers. Peers obtain the address of the tracker for a swarm by downloading a .torrent file from a website. Once a peer A has received a list of contacts in the swarm, it tries to establish connections to them. If it is successful, the according remote peer B is added to A's neighbor set and vice versa.

Thus, each overlay connection has one peer that initiated the connection, and one peer accepting it. In general, new peers fill their neighbor set by initiating most or all of their connections, while peers that have been online longer also have a significant share of connections they accepted. Since 'older' peers have more content and are therefore better sources for chunks, we want to test with our evaluation whether more data flows from peers accepting connections to the initiating ones than in the opposite direction.

Choke Algorithm. Every 10 seconds, a peer decides to which of its interested neighbors it will upload data to. These peers are called *unchoked*, the rest is choked. In standard BitTorrent, there are 3 regular unchoke slots which are awarded to the peers that offer the currently highest upload rate to the local peer. This strategy is called *tit-for-tat* and provides an incentive for peers to contribute upload bandwidth to the swarm. If the local peer has already downloaded the complete file, i.e., it is a *seeder*, the slots are given to all interested neighbors in a round-robin fashion.

Additionally, every 30 seconds a random peer not currently unchoked is selected for *optimistic unchoking* for the next 30 seconds. This allows a peer to discover new mutually beneficial data exchange connections.

Chunk Selection Algorithm. Once a peer is unchoked, it has to select which of the available chunks to download. BitTorrent uses the rarest-first strategy here, i.e., among all chunks the local peer still needs and the remote peer has to offer, the one that is seen the least in the neighbor set of the local peer is selected. This strategy tries to prevent chunks to be shared much less than others, risking the loss of chunks to the swarm.

2.2 Tribler

The VoD client of Tribler adapts the two most important mechanisms of BitTorrent, which are the peer selection strategy in the unchoking process and the piece or chunk selection strategy. We will describe them shortly here, since especially the changed chunk selection mechanism has consequences on the traffic exchange between peers. Details on these mechanisms can be found in [PGW+06].

Chunk Selection Algorithm. The main difference between a file-sharing functionality as offered by BitTorrent and a VoD service as offered by Tribler is that a user of the latter watches the video while downloading it. Thus, the timing for downloading the parts of the complete video file becomes critical, while chunks can be downloaded in any order in a file-sharing network. In particular, chunks in Tribler need to be downloaded roughly in order so that a continuous playback of the video can be ensured.

To this end, the rarest-first chunk selection of BitTorrent is replaced by a strategy based on priority sets. From the current playback position, all chunks until the end of the movie are separated into three sets. The high-priority set contains all chunks with frames from the playback position until 10 seconds after it, while the mid-priority set contains the following 40 seconds of the movie. The remainder comprises the low-priority set. Chunks are first downloaded from the high-priority set, following an in-order strategy within that set. Afterwards, the chunks in the mid-priority set are downloaded, and finally the chunks of the lowest priority, both according to the BitTorrent rarest-first mechanism.

Choke Algorithm. The tit-for-tat (T4T) strategy employed to rate peers in the unchoking process of BitTorrent is based on the assumption that peers can exchange content, i.e., both neighbors forming a connection have some content the other needs. Since the order in which peers download chunks does not matter in the file-sharing application, this is true for enough overlay neighbors. However, with the application VoD peers need to download chunks in roughly the order they are played out as explained above.

Therefore, peers that played back a longer part of the video do generally not need any chunks from peers that are 'behind' them in the playback process. Thus, it is much more probable that data exchange happens only in one direction of an overlay connection in Tribler, namely from a peer that is online longer to a peer that has joined later.

This is taken into account by replacing T4T with a strategy named give-to-get (G2G). G2G favors peers that show a good upload behavior to other peers instead of to the local peer. Consider the local peer A, which has chunks a remote neighbor B wants to download. B then reports to which other peers it has uploaded data in the last δ seconds (by default, $\delta = 10$). Then A queries these peers to make sure that B does not exploit the mechanism. The rating value of B then is the amount of data it forwarded that it originally received from A. Only if there is a tie using this metric, B's total upload is considered as a tie-breaker. This ensures that peers with a high upload capacity are not unchoked by a large number of neighbors, but only by a selected subset.

2.3   Relevance for Economic Traffic Management

ETM approaches utilizing locality-awareness, such as BNS or BU, influence the neighbor selection and the unchoking process of clients. The former method tries to establish connections to close peers, e.g., peers in the same AS. This is mainly achieved by changing the decision to which other peer a local peer will initiate a connection to. If not all peers implement such an ETM mechanism, and if more traffic flows into one direction of a connection than in the other, this leads to an unbalanced effect on the traffic flows of the peers promoting locality, as we show in our results.

On the other hand, BU prefers local neighbors when optimistically unchoking another peer. Thus, it influences the upload of a peer more than its download. Still, due to the T4T algorithm, it has an indirect effect on the download of a peer as well. Another method of relevance here would be the restriction of signaling interest to only a subset of peers instead of all peers holding missing content. However, we will not consider this mechanism here.

2.4   Related Work

A number of measurement and analytical studies on BitTorrent swarms can be found in literature. In [IUKB+04], tracker logs of a 5 month period are evaluated for one swarm. Additionally, a modified client is used to gather data. This work gives some important information about client behavior and swarm dynamics for the BitTorrent protocol current at the time of its publication. While there is no explicit differentiation between overlay connection types of clients, first conclusions are drawn about the composition of a peer's neighbor set. It is stated that a neighbor set contains 'old' as well as 'new' peers, enabling every peer to find some sources for data. This can be mapped to our considerations about initiated connections (to 'older' peers) and accepted ones (from 'newer' peers). Thus, we provide a more in-depth view of this special phenomenon, and additionally evaluate its effect on ETM.

Another trace-based analysis of BitTorrent swarms is provided by [PGES05]. An index site as well as the associated trackers are monitored over a period of 8 months in 2004, with a focus on the availability of system components. This includes the infrastructure that is not part of the core BitTorrent protocol itself. The results cover important aspects like peer uptime, swarm size over time and average download speeds, but take no detailed look at overlay structures.

A more recent and comprehensive measurement study of BitTorrent swarms was conducted in [HHO+09]. Here, data from more than 300,000 swarms active between June 2008 and May 2009 was gathered and evaluated, in addition to existing data sets. Among other things, it was found that swarm characteristics depend strongly on the type of content shared. One of the most important results for this work is the skewness of the peer distribution in the network topology, something which we model in our scenarios.

In [BHP06], a simulative approach to analyze the mechanisms of the BitTorrent protocol is taken. The focus here is on the efficiency of the rarest-first chunk selection and of the tit-for-tat unchoking policy. The study concludes that the load in a swarm is distributed unfairly, since peers with a higher capacity upload more than their fair share. However, this seems to be only a minor problem in active swarms. Nevertheless, mechanisms are presented that limit this unfairness at the cost of protocol efficiency. Additionally, experiments are conducted where nodes join a swarm in the post-flash crowd phase, i.e., when new nodes with little or no content connect to older nodes with many chunks. Similarly, 'pre-seeded' nodes, i.e., nodes having completed most of the file download, are added to a swarm during its initial flash-crowd phase.

In the former case, it is shown that the new nodes quickly download chunks that are interesting to their neighbors, enabling them to participate in the swarm. Here, outgoing connections are considered, but no comparison to incoming connections is made. In the case of pre-seeded nodes joining a swarm, the results show that these nodes might take much longer to complete their download than in a scenario where all nodes have the same share of the content. However, the number of neighbors used in the evaluation scenarios is much lower than in a real swarm with the current version of the protocol, which might influence the results.

Finally, an analytical model for BitTorrent swarms was presented in [QS04]. It captures values like the leecher and seeder population of a swarm and its upload and download capacity to draw conclusions about the evolution of the peer population over time and about the average download time. By necessity, the level of abstraction of the model prohibits taking into account mechanism details such as the neighbor selection or unchoking mechanisms.

None of the works above distinguishes in detail between different connection types in the neighbor set of a peer. To the best of our knowledge, there is no work considering the correlation between traffic flows and the type of connection between two neighbors. Moreover, no such study exists for current approaches to ETM.

3 Simulation Model

Our evaluation is based on a event-driven simulation. In the following, we describe our default simulation scenario and the simulator used. Specific parameters that are changed in the experiments are described in Section 4.

3.1 Simulation Scenario

We simulate one BitTorrent or Tribler swarm which exchanges a file of size 154.6 MB generated from an example TV show of about 21 minutes in medium quality. The file is divided into chunks of 512 KB and every chunk into blocks of 16 KB. In case of Tribler, these values translate to a stream bitrate of slightly below 1 Mbit/s.

We simulate the swarm for 6.5 hours, of which it is in the steady state for 5 hours. New peers join the swarm with an exponentially distributed inter-arrival time A with a mean value of $E[A] = 10$ s. As a result, one simulation run consists of about 2300 downloads in the default scenario. The peers stay online for the full download duration of the file plus an additional, exponentially distributed seeding time with a mean value of 10 minutes. Peers do not go offline during the download or the seeding time. As a result, we measured that the swarm contains on average about 120 peers depending on the scenario. These parameters are the same as in [OLH+09].

The peers are connected with an access speed of 16 Mbps downstream and 1 Mbps upstream, which are typical values for the DSL access technology. The seed has a symmetric upload and download bandwidth of 10 Mbps. It goes offline after 1 hour of simulation time. We model the inter-AS links as well dimensioned, i.e., the only bottlenecks are the access links of the peers.

3.2 Simulator

The simulator used in this work is a current version of the simulator used in [OLH+09]. It is based on the P2P simulation and prototyping Java framework ProtoPeer [pro, GADK09]. The simulator contains a flow-based network model adopting the max-min-fair-share principle [BG87]. It faithfully implements the BitTorrent functionality and behavior as described in [LUKM06] and [Bit]. It includes all key mechanisms, in particular the piece selection mechanisms, the management of the neighbor set, and the choke algorithm. Furthermore, the complete message exchange among the peers themselves, between peers and the tracker as well as between the peers and the information service for locality data is simulated in detail. For more specifics, we refer to [OLH+09].

4 Experiments and Results

In this section, we present the results from our simulative performance evaluation. For the measured traffic amounts, we average over all peers in one simulation run, and show the mean values and 95% confidence intervals over 10 runs.

In the following experiments, we vary the seeding time and the interarrival time of the peers and compare the BitTorrent protocol with the Tribler protocol. Additionally, we take a look at the upload and download traffic of an ISP implementing ETM.

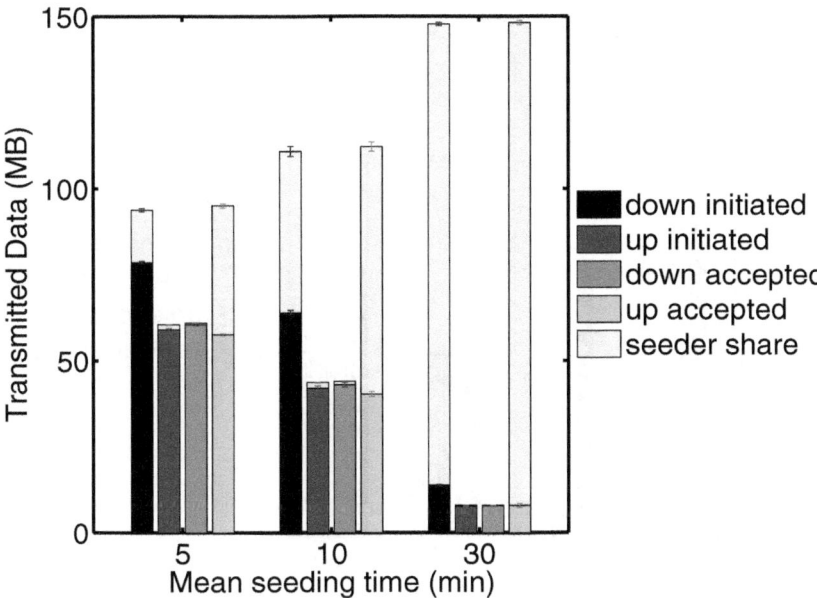

Fig. 1. Traffic direction for different seeding times

4.1 Seeding Time

In the first experiment, we let the peers go offline after mean seeding times of 5, 10 and 30 minutes, cf. Figure 1. Each group of bars shows the average volume of data per peer downloaded and uploaded over initiated and accepted connections, respectively. The light blue bars on top are the share of data that was downloaded from seeders (in case of download traffic), and uploaded while being in seeder mode (in case of upload traffic).

Our first observation is that more traffic is downloaded via initiated connections than via accepted ones, and that similarly, more traffic is uploaded via accepted connections. A large share of this traffic flows from seeders to leechers, although even among leechers, initiated connections are used more heavily for download than accepted ones.

This biased direction of traffic flows is important when trying to influence P2P traffic flows by changing overlay connections, since a mechanism changing the selection of neighbors on the initiator side mainly affects traffic flowing in the opposite direction.

For higher seeding times, the discrepancy between initiated and accepted connections grows. However, this is mainly due to the larger share of traffic uploaded from seeders, which are contacted more often from leechers than the other way around. Thus, depending on the seeding behavior of peers, an ETM changing client behavior should not only work in the leeching mode, but also when the client in question is a seeder.

4.2 Interarrival Time

In this experiment, we test whether the connection usage changes for different swarm sizes and a different peer arrival rate. We compare Poisson arrival processes with a

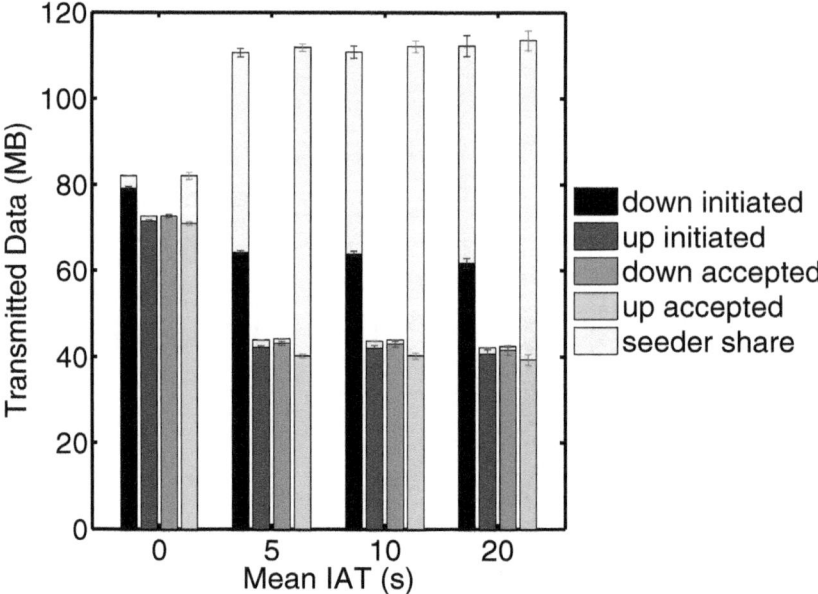

Fig. 2. Traffic direction for different IATs

mean interarrival time of 5, 10 and 20 seconds. Additionally, we evaluate a scenario where 500 peers join the swarm at the same time, i.e., a flash-crowd scenario (denoted by 'Mean IAT 0'). The results are shown in Figure 2.

We observe that, for the swarms in the steady state, the interarrival time and the resulting change in the peer population leads to no significant differences in the connection usage. The effects are the same as explained above. However, this changes for the flash-crowd scenario. Here, the peers all start downloading the file form the initial seed at the same time, and therefore no additional seeders exist in the swarm during most of their leecher phase. Additionally, peers show a similar download progress and establish connections to each other at roughly the same time. As a result, the difference between data flowing over initiated and accepted connections is much smaller than in the steady state scenarios.

4.3 Overlay Type

Next, we compare the predominant BitTorrent protocol with Tribler to see whether the application of VoD streaming changes the distribution of traffic over the different types of overlay connection. To be able to fairly compare the two protocols, we use a fixed deterministic online time of 25 min for all peers. This is long enough to ensure that all peers download or play out the complete file, but is independent from their download mechanisms, in contrast to seeding times.

As discussed in Section 2.2, the in-order download of chunks should lead to more uni-directional traffic flows in the overlay. This can be observed in Figure 3. The

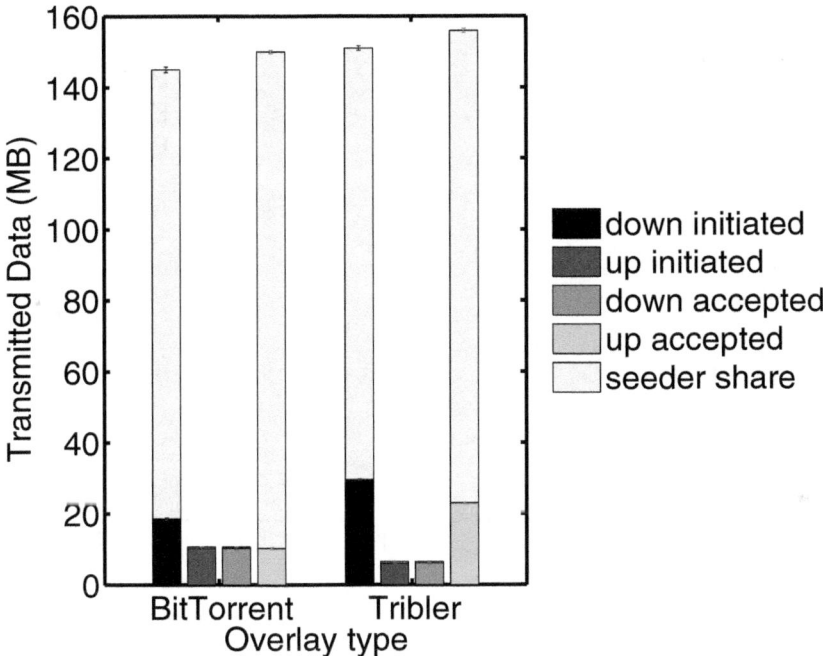

Fig. 3. Traffic direction for different overlay types, BitTorrent and Tribler

difference in traffic flowing over initiated and accepted connections is more pronounced in Tribler, with more traffic flowing to the peers that initiate connections.

Thus, an ETM mechanism that wants to be efficient not only for file-sharing, but also for the future application of video streaming, needs to take into account the inherent hierarchy among peers sharing the same video. A mechanism that governs mainly to which peers connections are initiated influences even more the sources of traffic flowing to the local peer, and even less the direction of upstream traffic flows.

4.4   Effect on Inter-AS Traffic

Until now, we have considered the traffic flows per peer in a regular BitTorrent swarm. To judge the effect ETM mechanisms have, we next take a look at the traffic savings using two different approaches, BNS and BU. In this scenario, we use a swarm setup based on measurements of live BitTorrent swarms [HHO+09]. We use a topology with 20 ASes, and set the probability of a peer going online in AS k is $P(k) = \frac{\frac{1}{k}}{\sum_{i=1}^{20} \frac{1}{i}}$. This models the skewed peer distribution observed in real swarms. All peers use the default BitTorrent protocol, except for the peers in the largest AS 1. These implement the described locality-awareness mechanisms, BNS, BU, and the combination of both. This way, we show the effects on the traffic resulting from a part of the swarm implementing ETM.

We measure both the incoming and the outgoing traffic for each AS. Figure 4 shows the incoming traffic for the four different client implementations in AS1. We observe

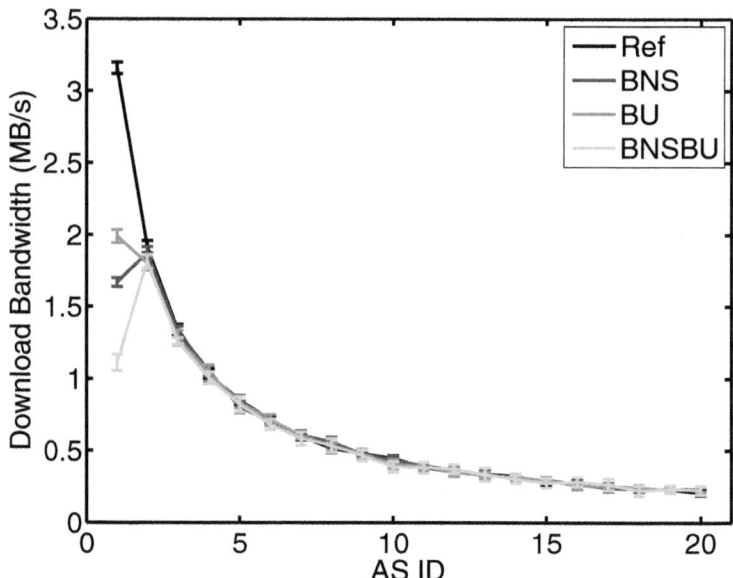

Fig. 4. Download traffic for different ETM mechanisms

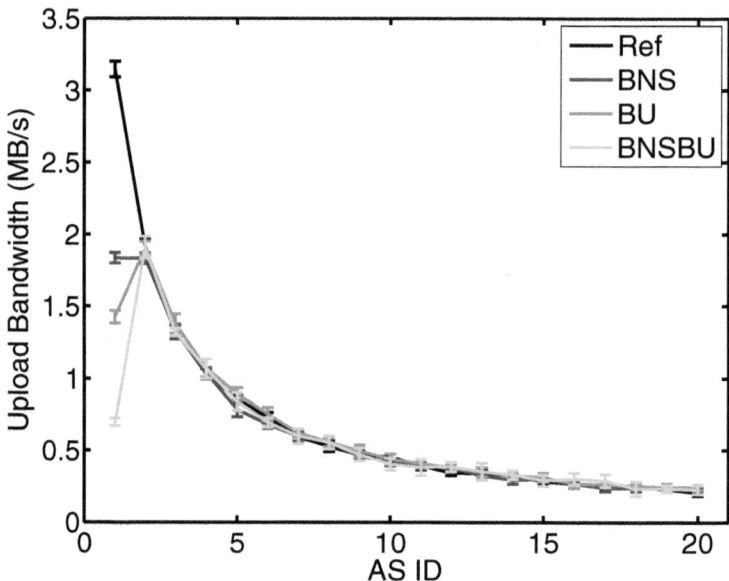

Fig. 5. Upload traffic for different ETM mechanisms

large differences in the effectiveness of the mechanisms, which can be explained by the results above. BNS governs mainly how connections are initiated. In conjunction with the fact that initiated connections carry more incoming traffic, this leads to less consumed download bandwidth in comparison to BU, which influences only the upload of a peer. Still, BU saves traffic due to the tit-for-tat algorithm and because the other peers in AS1 also upload preferredly to local peers.

In contrast, BU proves to be more effective than BNS in reducing the upload of AS1, cf. Figure 5. BNS reduces the upload traffic less, both in comparison to BU and in comparison to the download traffic. This again is due to the biased usage of overlay connections. The combination of both mechanisms reduces the upload traffic much stronger than the download traffic, which can be attributed to the fact that BNS supports BU very well in its preference to upload to local peers, but not the other way round.

5 Conclusions

The results presented in this work show that more traffic flows from peers accepting an overlay connection to the peer initiating it. This effect is enhanced when a peer becomes a seeder, since then it is only an uploader which accepts much more connections than it initiates. Additionally, the amount of traffic flowing from seeders to leechers is significant even for short seeding times. Therefore, we conclude that any efficient ETM has to take this fact into account. Influencing the peer neighbor selection behavior on the initiator's side influences the traffic flows towards the local peer stronger than the opposite direction. Moreover, the ETM mechanism should be as efficient in the seeder mode as in the leecher mode, such as Biased Unchoking using the optimistic unchoke slot, which is utilized in both states. We showed that the described effect is observable in a realistic swarm where only a part of the overlay employs ETM.

Acknowledgments

This work has been performed in the framework of the EU ICT Project SmoothIT (FP7-2007-ICT-216259). The authors would like to thank Tobias Hoßfeld in particular for his help with conducting runs.

References

[BCC+06]   Bindal, R., Cao, P., Chan, W., Medval, J., Suwala, G., Bates, T., Zhang, A.: Improving traffic locality in bittorrent via biased neighbor selection. In: Proceedings of the 26th IEEE International Conference on Distributed Computing Systems, p. 66. IEEE Computer Society, Washington (2006)

[BG87]   Bertsekas, D., Gallagher, R.: Data Networks. Prentice-Hall, Englewood Cliffs (1987)

[BHP06]   Bharambe, A.R., Herley, C., Padmanabhan, V.N.: Analyzing and improving a bittorrent networks performance mechanisms. In: 25th IEEE International Conference on Computer Communications (INFOCOM 2006), pp. 1–12 (April 2006)

[Bit]   Bittorrent specification,
               http://wiki.theory.org/BitTorrentSpecification
[CS09]   Inc. Cisco Systems. Cisco Visual Networking Index: Forecast and Methodology,
               2008-2013. White Paper (June 2009)
[GADK09]   Galuba, W., Aberer, K., Despotovic, Z., Kellerer, W.: ProtoPeer: A P2P Toolkit
               Bridging the Gap Between Simulation and Live Deployment. In: Proceedings of
               the 2nd International Conference on Simulation Tools and Techniques (2009)
[GRH+08]   Gimenez, J.P.F.-P., Rodriguez, M.A.C., Hasan, H., Hoßfeld, T., Staehle, D., Despo-
               tovic, Z., Kellerer, W., Pussep, K., Papafili, I., Stamoulis, G.D., Stiller, B.: A new
               approach for managing traffic of overlay applications of the smoothIT project. In:
               Hausheer, D., Schönwälder, J. (eds.) AIMS 2008. LNCS, vol. 5127, Springer, Hei-
               delberg (2008)
[HHO+09]   Hoßfeld, T., Hock, D., Oechsner, S., Lehrieder, F., Despotovic, Z., Kellerer, W.,
               Michel, M.: Measurement of bittorrent swarms and their as topologies. Technical
               Report 463, University of Würzburg (November 2009)
[IUKB+04]   Izal, M., Urvoy-Keller, G., Biersack, E.W., Felber, P.A., Al, Garcés-Erice, L.: Dis-
               secting bittorrent: Five months in a torrent's lifetim, pp. 1–11 (2004)
[LUKM06]   Legout, A., Urvoy-Keller, G., Michiardi, P.: Rarest first and choke algorithms are
               enough (2006)
[OLH+09]   Oechsner, S., Lehrieder, F., Hoßfeld, T., Metzger, F., Pussep, K., Staehle, D.: Push-
               ing the performance of biased neighbor selection through biased unchoking. In:
               9th International Conference on Peer-to-Peer Computing, Seattle, USA (Septem-
               ber 2009)
[OSP+08]   Oechsner, S., Soursos, S., Papafili, I., Hoßfeld, T., Stamoulis, G.D., Stiller, B.,
               Callejo, M.A., Staehle, D.: A Framework of Economic Traffic Management Em-
               ploying Self-organization Overlay Mechanisms. In: Hummel, K.A., Sterbenz,
               J.P.G. (eds.) IWSOS 2008. LNCS, vol. 5343, pp. 84–96. Springer, Heidelberg
               (2008)
[PGES05]   Pouwelse, J.A., Garbacki, P., Epema, D.H.J., Sips, H.J.: The bittorrent p2p file-
               sharing system: Measurements and analysis. In: Castro, M., van Renesse, R. (eds.)
               IPTPS 2005. LNCS, vol. 3640, pp. 205–216. Springer, Heidelberg (2005)
[PGW+06]   Pouwelse, J.A., Garbacki, P., Wang, J., Bakker, A., Yang, J., Iosup, A., Epema,
               D.H.J., Reinders, M., Van Steen, M.R., Sips, H.J.: Tribler: A social-based peer-to-
               peer system. In: The 5th International Workshop on Peer-to-Peer Systems (IPTPS
               2006), pp. 1–6 (2006)
[pro]   Protopeer, http://protopeer.epfl.ch/index.html
[QS04]   Qiu, D., Srikant, R.: Modeling and performance analysis of bittorrent-like peer-
               to-peer networks. In: SIGCOMM 2004: Proceedings of the 2004 Conference on
               Applications, Technologies, Architectures, and Protocols for Computer Communi-
               cations, pp. 367–378. ACM, New York (2004)

Implementation and Performance Evaluation of the re-ECN Protocol

Mirja Kühlewind[1] and Michael Scharf[2],[*]

[1] Institute of Communication Networks and Computer Engineering (IKR)
University of Stuttgart, Germany
mirja.kuehlewind@ikr.uni-stuttgart.de
[2] Alcatel-Lucent Bell Labs, Stuttgart, Germany
michael.scharf@alcatel-lucent.com

Abstract. Re-inserted ECN (re-ECN) is a proposed TCP/IP extension that informs the routers on a path about the estimated level of congestion. The re-ECN protocol extends the Explicit Congestion Notification (ECN) mechanism and reinserts the obtained feedback into the network. This exposure of congestion information is a new economic traffic management mechanism that enables the network to share the available capacity more equally and to police the compliance of congestion control through e. g. a per-user congestion limitation.

This paper studies performance implications of the re-ECN mechanism. Our evaluation is based on simulations with an own re-ECN implementation in the Linux TCP/IP stack. Our results also confirm that congestion exposure generally works. But we also show that traffic characteristics such as the round-trip time (RTT), flow sizes, as well as the selection of congestion control algorithms have a significant impact on the congestion exposure information. These non-trivial effects have to be taken into account when using the re-ECN information as input parameter for congestion control mechanisms in end-systems or for routing/policing inside the network. Comparable results have not been published so far.

1 Introduction

In the Internet, the congestion control mechanisms of the Transmission Control Protocol (TCP) mainly decide how resources are shared [1]. Today's practice does not ensure a fair share of the available bandwidth in respect to all kind of emerging application requirements. This problem becomes increasingly critical as peer-to-peer file sharing and video streaming services consume already a large part of the available bandwidth with a strong upwards trend. In addition, more and more inelastic traffic emerges (e. g., television). A further challenge for resource sharing are new high-speed congestion control mechanisms (e. g., [2]) that are more aggressive than TCP's standard congestion control algorithms.

As a response, more and more network providers use e. g. Deep Packet Inspection in combination with traffic shapers to penalize bulk data applications and

---

[*] The major part of this work was performed while the author was with IKR.

B. Stiller, T. Hoßfeld, and G.D. Stamoulis (Eds.): ETM 2010, LNCS 6236, pp. 39–50, 2010.

users that cause large traffic volumes [3]. But these approaches do not consider the actual congestion situation on the path, nor do they provide incentives for the users to react appropriately. A promising alternative is an accountability system that encourages the end-systems to share the bandwidth fairly. That means the available bandwidth should be used efficiently if no congestion occurs. But at the same time, time-critical applications should not significantly be affected even in a congestion situation. One solution to make end-systems accountable for their impact on the network is to expose the level of expected congestion on a network path to the network components on the path. This can be achieved thought the re-ECN protocol which has been proposed by Briscoe *et al.* [4]. It extends the Explicit Congestion Notification (ECN) mechanism [5] and reinserts the obtained feedback in the network, using the principle of re-feedback [6]. It is an incentive-oriented, end-to-end protocol that also enables congestion-based policing in the network, which has a significant potential for economic traffic management [7]. re-ECN is currently discussed in the Internet Engineering Task Force (IETF) as a solution for the congestion exposure problem [8].

The key contribution of this paper is an independent evaluation of the re-ECN protocol. So far there are only a few studies of Briscoe *et al.* [6]. In order to verify the specification, we implemented the protocol for the Internet Protocol version 4 (IPv4) and TCP in the Linux network stack. Our implementation is the only kernel implementation besides the one of Alan Smith used by Briscoe *et al.* The use of real kernel code with the Network Simulation Cradle (NSC) framework [9] allows us to perform realistic user-space simulations.

As this is an initial evaluation, our focus is to highlight the principle operation of the re-ECN protocol. Our results reveal several important characteristics of re-ECN that affect the performance of the proposed congestion-based traffic management: The congestion exposure information depends on the RTT, on the selected congestion control mechanism, and/or on the parameterization of the Active Queue Management (AQM) in a router. Furthermore, the TCP Slow-Start overshoot can cause congestion peaks that could discriminate flows with a certain length. These non-trivial dependencies have to be taken into account when using re-ECN information for economic traffic management. In the long term re-ECN is intented to encourage the use of transport protocols that adapt to the users' intentions.

The remainder of this paper is structured as follows: Section 2 gives an overview of the re-ECN protocol. In Section 3, we highlight some of the implementation challenges, and we also outline some open issues in the specification. The setup of our simulations is explained in Section 4. Section 5 then presents selected results for the performance of re-ECN. Section 6 concludes the paper.

2   Overview of re-ECN

2.1   The re-ECN Protocol for TCP/IPv4

re-ECN is an extension of ECN [5], which signals congestion to the sender before packets get lost. Like ECN, it requires routers to use AQM, such as Random

Early Detection (RED) [10], to mark a packet instead of dropping it. Full re-ECN support requires modifications in the receiver as well (RECN mode). If the receiver is only ECN capable, re-ECN will operate in RECN-Co mode.

In order to expose the expected congestion to all components on a network path, the sender uses bit 48 of the IPv4 header, the so called re-Echo (RE) flag. The receiver counts the packets that have been marked as Congestion Experienced (CE) by an ECN router and returns this value back to the sender. For every congestion announcement the sender has to blank the RE flag of a data packet. Otherwise, the RE flag is always set to "1". The fraction of re-ECN capable packets with the RE flag blanked can be considered as an estimation of the expected congestion on the path.

The re-ECN extension flag together with the two-bit ECN IPv4 field provide 8 available codepoints shown in Table 1. The encoding conserves backwards compatibility to ECN. The Feedback Not Established (FNE) codepoint should be used in the TCP handshake and at the beginning of a data transmission or after an idle time when not sufficient feedback from the receiver is available. In [4] it is recommended to set the first and third data packet of a re-ECN transmission to FNE. Pure ACKs, re-transmissions, window probes, and partial ACKs are supposed to be Not-ECT marked.

Table 1. Description of the extended re-ECN codepoints

ECN field	RE flag	RFC3168 codepoint	re-ECN codepoint	Worth	Description
00	0	Not-ECT	Not-ECT	–	Not re-ECN capable
00	1	-	FNE	+1	Feedback not established
01	0	ECT(1)	Re-Echo	+1	Re-echoed congestion and RECT
01	1	-	RECT	0	re-ECN capable
10	0	ECT(0)	ECT(0)	–	ECN use only
10	1	-	–	–	unused
11	0	CE	CE(0)	0	Re-Echo cancelled
11	1	-	CE(-1)	-1	Congestion experienced

Table 1 also displays a worth for every re-ECN codepoint. Packets with the RE flag blanked, as well as FNE packets, can be considered as credits for congestion a sender might cause on the path. The FNE codepoint should provide initial credits when no feedback has been observed so far. In contrast, every CE marking of a router counts as debit. As the CE(0) codepoint has the RE flag blanked and is CE marked, it is neutral.

2.2   Overall re-ECN Framework

A re-ECN framework with a policer at network ingress and a so-called dropper at network egress is proposed in [6]. The bottom part of Fig. 1 illustrates the congestion level signaling with re-ECN. The fraction of positive (RE blanked or FNE) packets, illustrated in black, is the whole-path congestion level declared

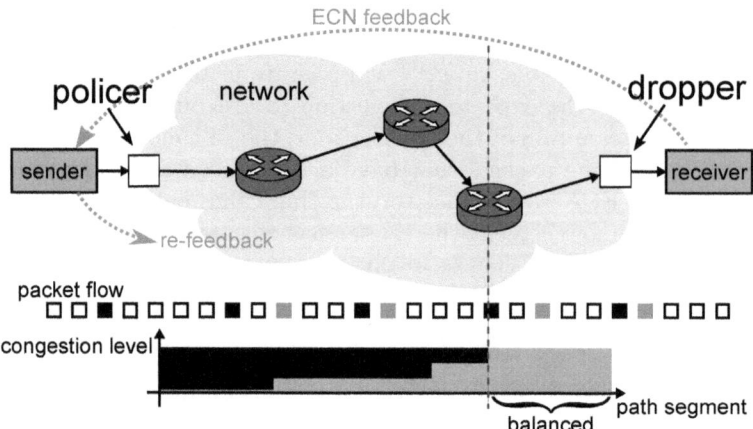

Fig. 1. re-ECN framework and the congestion level resulting from the balance of marks

by the sender. The fraction of negative (CE marked) packets, in gray, increases over the network path as more and more packets get marked by congested routers. At the end of the path the rate of black and gray packets should be about the same if the sender declares honestly the expected congestion level. Hence, the difference between the fraction of positive (Re-Echo or FNE) and negative (CE(-1)) marks of all packets gives the expected congestion level for the remaining path at every point in the network.

The re-ECN framework sketched in Fig. 1 requires a dropper component at network egress to ensure that the sender honestly declares its expected whole-path congestion. If the number of negative marked packets is larger than the number of positive marked ones, the dropper can detect that a sender underestimates its congestion and subsequently penalize its flow through packet drops.

At the same time a policer at network ingress can restrict how much resources one end-system can allocate on congested links, based on end-system's whole-path congestion estimation. Such a policer encourages senders to only cause congestion if really needed or to decrease the sending rate instead. This can for example be implemented through a token bucket mechanism with a certain allowed congestion rate and maximum congestion volume per end-system. If the given limits are exceeded, the policer will enforce the restrictions, e. g., through packet drops. As this paper focuses on the signaling mechanism of re-ECN, the design and implications of droppers and policers are left for further study.

The re-ECN framework provides incentives to end-systems to perform appropriate congestion control. In particular, it creates a motivation to use less than best effort congestion control schemes [11] for bulk data transfers that are not time-critical. Furthermore, it could be one component of a disruptive new Internet capacity sharing architecture that uses new congestion control algorithms not designed to be TCP-friendly (e. g., [12]).

3   re-ECN Implementation

3.1   Implementation of re-ECN in Linux

We have implemented re-ECN for IPv4 and TCP in the Linux kernel 2.6.26. Our implementation conforms to the protocol specification in [4], except that it does not send re-ECN marks after lost packets, which is not required in our studies since all components are ECN capable. In our implementation the CE marks are counted on a per-packet base. In contrast, the implementation of Briscoe *et al.* uses a per-segment solution. Our implementation is completely independent from this code. As all ECN functionality is implemented in separate methods in the Linux kernel, most of the re-ECN processing was simple to realize.

3.2   Lessons Learned

During the development of our re-ECN implementation, we found issues that are not addressed by the existing specification. First, the re-ECN specification does not address how to deal with fragmentation of IP packets. A router that has to fragment IP packets does not know how to set the markings correctly. As a remedy, in our implementation every fragmented IP packet is set to Not-ECT. A similar problem occurs for Generic Segmentation Offload (GSO) which would require the transfer of TCP state information. In addition, the specification [4] recommands to set the first and thrid packet to FNE. This is based on the assumption that the level of congestion is constant, which is not correct. The problems that result form this recommandation are explained in Section 5.2.

A number of specific design choices of the Linux TCP/IP stack also complicates the implementation of re-ECN: SYN cookies as implemented in the Linux kernel cannot be used with ECN and therefore with re-ECN neither. The check of re-ECN information in our implementation is supported by both, the fast and slow path processing of TCP ACKs in the Linux kernel. Slow path processing only would result in significant delays of some positive markings. Furthermore, the Linux stack is not well prepared to handle a new usage of the unused bit 48 in the IP header. Finally, the implementation of the RECN-Co mode is non-trivial, because there is a problem with tracing the transition from '1' to '0' of the ECN Expected Congestion Echo (ECE) flag in the Linux kernel implementation. Anyway, the RECN-Co mode is not able to provide feedback on more than one observed congestion event per RTT.

4   Simulation Setup

4.1   Simulation Tool

In order to investigate the re-ECN performance, the TCP/IP network stack of the Linux kernel was used within a simulation environment. The use of operating system code more realistically models the behavior of a TCP/IP stack, whereas abstract TCP models in simulation tools are often overly simplified. Our tool is based on the IKR Simulation Library [13], and the NSC [9] is used to integrate the kernel code within simulation components.

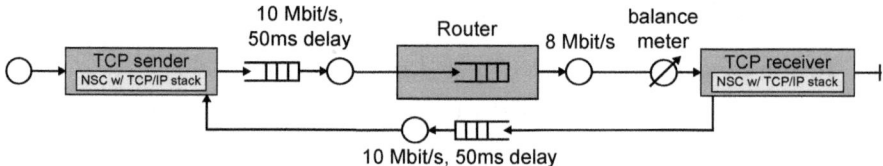

Fig. 2. Simple re-ECN simulation scenario with one TCP sender

4.2 Evaluation Scenario

For a first evaluation of the implications of the re-ECN protocol, a simulation scenario was used with just two end-systems both supporting re-ECN. The setup is displayed in Fig. 2. In the simplest case, one TCP endpoint sends bulk data traffic generated by a greedy source. Alternatively, we use a burst traffic generator with on/off traffic in order to evaluate the impact of small flows on the congestion exposure with re-ECN. We also use simulations with constant bit rate (CBR) or TCP cross traffic. Finally, to emulate a realistic workload scenario [14], we replay Internet traffic traces that are provided in [15]. The flows then start with an negative exponential distributed inter-arrival time of 100 ms.

In the simple setup, the first link between sender and router has a link capacity of 10 Mbit/s and a delay of 50 ms. The second link between router and receiver provides just 8 Mbit/s and is thus the bottleneck. The reverse path also has a capacity of 10 Mbit/s and a delay of 50 ms. This scenario reveals how the information exposed by re-ECN depends on the AQM mechanism. When replaying the real Internet traces, all transmissions share one bottleneck link of 10 Mbit/s in both directions and have an access bandwidth of 100 Mbit/s. We use 9 different RTTs, as recommended in [14].

In every scenario the router on the path uses RED with a queue length of 100 packets. RED's minimum threshold to mark packets is 10 packets. If the average queue size exceeds this threshold, packets are marked with a certain increasing probability. There is a maximum (hard mark) threshold to mark every packet if the average queue size is above 50 packets. The RED algorithm requires a weight factor to calculate the average queue size, which is always set to 1/32 in our simulations. As our RED implementation is based on the Linux kernel code, there is a parameter for the number of random bits which determines the maximum mark probability. This parameter is always set to 9.

In order to study the performance of re-ECN, a meter in front of every receiver counts the total number of negative and positive marked packets. The difference gives the current re-ECN balance. At network egress it should usually be zero as explained in Sect. 2.2. In our calculation we always measure all observed marks from the beginning of the transfer. As there are three FNE packets (SYN, first and third data packet) at the beginning of every data transmission, the rest-of-path congestion is balanced with a value of 3, instead of 0. An alternative to determine the balance would have been a weighted moving average. But for our studies the actual total balance reveals more precise information.

5 Performance Results

5.1 Principle Behavior of re-ECN

As a first step, the fundamental behavior of the re-ECN protocol is evaluated. For this purpose we use the simple scenario described in Sect. 4.2 with one TCP bulk data sender. The sender honestly exposes its observed congestion and can use different congestion control mechanisms. Fig. 3 plots both the congestion window (CWND) size and the re-ECN balance measured by the meter if the sender either uses the Reno [1] or the CUBIC [2] congestion control algorithm.

The diagrams in Fig. 3 show TCP's typical sawtooth pattern in the CWND evolution over time. In the trace of the re-ECN balance, several aspects can be observed: First, the balance is 3 most of the time due to the initial FNE marks. Second, each time packets get marked by the RED queue, there is a negative peak in the re-ECN balance, which is caused by delayed re-insertion of the re-ECN information. The peak size depends on the number of marked

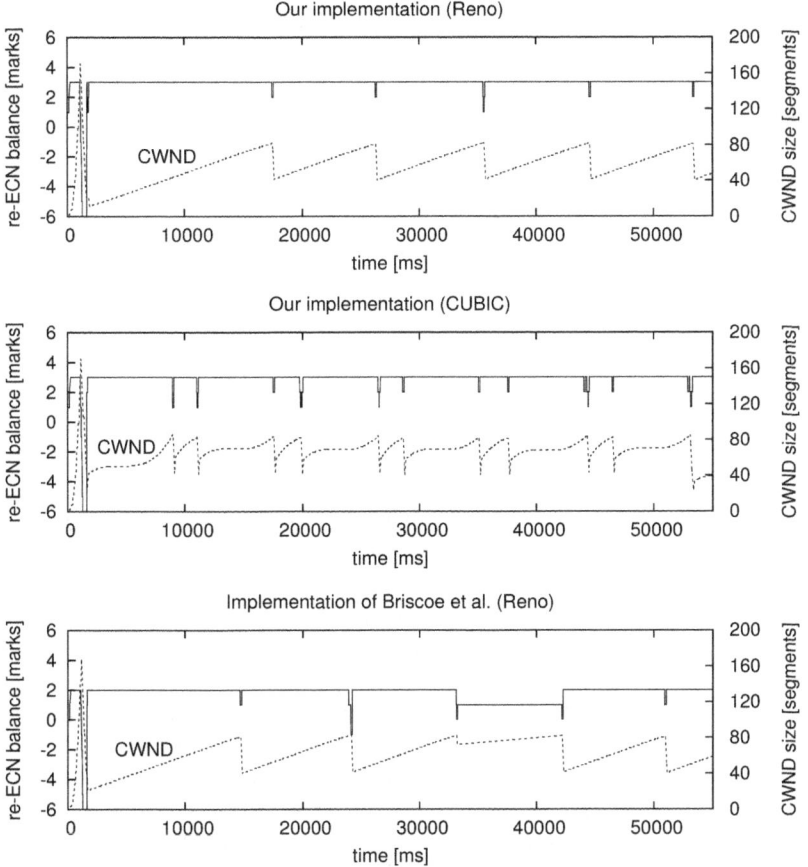

Fig. 3. re-ECN balance of a TCP connection at egress

packets. In all diagrams, there is a significant first negative peak of about -60 in the re-ECN balance, which occurs a couple of RTTs after the connection setup. This is caused by the well-known overshoot effect in TCP Slow-Start. This effect needs to be considered in the design of droppers and might result in a need to identify flow starts. Thereby, the dropper design may get more complicated and the potential for attacks increases. In the long term re-ECN is indented to foster congestion-improved start-up mechanisms.

It must be noted that it takes more than one RTT after the first negative marks until the re-ECN balance is recovered. One RTT is already needed to transport the congestion exposure information. Moreover, the congestion window and respectively the sending rate get reduced as soon as the sender receives the congestion signal. In the examples in Fig. 3 having a one-way delay of 50 ms, the recovery of balance after the first negative peak takes 766 ms for Reno and 622 ms for CUBIC, since several CE marking have appeared in a row.

The first two diagrams in Fig. 3 also show that the amount of congestion depends on the congestion control algorithm of the sender. If the CWND is increased more aggressively, the number of marked packets increases, too. re-ECN correctly exposes the higher congestion level of a more aggressive sender.

The third diagram displays the behavior of the re-ECN implementation of Briscoe et al., using the same simulation scenario. The general behavior is similar. However, the implementation of Briscoe et al. results in an incorrect re-ECN balance between two congestion events. This problem seems to be caused by the lack of checking the re-ECN TCP flags in the fast path processing.

5.2   re-ECN Dependencies on Path and Traffic Charateristics

The presented results show that the re-ECN protocol in principle works and correctly informs all network components on the path about the amount of congestion, which would otherwise be known by the sender only. Still, the re-ECN balance will be incorrect if positively marked packets get lost or negatively marked ones get dropped in the receiver, e. g., because of a checksum error. A loss of four subsequent ACKs will cause a mismatch of 8 between the receiver and sender, as the 3 bit counter then may wrap around. Furthermore, the exposure information is at least one RTT delayed, which can result in a permanent negative offset when a congestion situation persists.

Fig. 4 shows the mean negative peak size, apart from the first negative peak in Slow-Start, as a function of the one-way delay, where a negative peak is a temporarily negative balance as to be observed in Fig. 3. The diagram reveals that the latency influences the number of re-ECN markings if the sender uses CUBIC. The larger the RTT of a flow, the larger is the temporary mismatch of the balance. As a result, it does not make sense to police a flow if its re-ECN balance exceeds a fixed negative threshold. Instead, a dropper has to observe a flow over a certain time period to ensure that the balance will not recover. However, the recovery time depends again on the RTT, the congestion control, as well as the number of markings, which complicates dropper design.

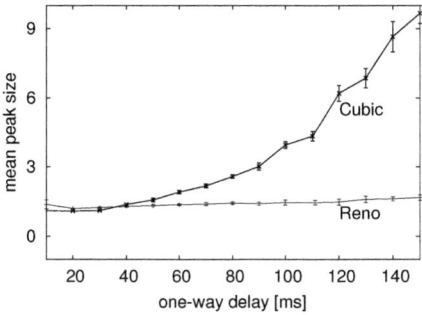

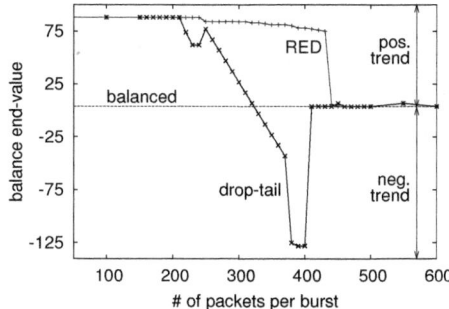

Fig. 4. Mean negative peak size with different one-way delays

Fig. 5. Balance after 1 minute for burst traffic with variable burst length

The size of a data transfer can also have an impact on the re-ECN balance. In order to illustrate this, we considered traffic of several data bursts with an idle time of about 2 s in between. We also compared the default RED configuration in this paper with an alternative setup, which has a hard mark limit at 10 packets. The latter case emulates a drop-tail marking behavior. Fig. 5 prints the re-ECN balance end-value after 1 minute. The diagram shows two effects.

First, if the burst length is rather short, the total balance of a flow has a large positive value. The re-ECN specification mandates that after an idle period of 1 s some data packets should be FNE marked as the congestion information in the sender is expired. Thus, each burst increases the balance. Such a bursty traffic will consume a lot of the congestion volume in a policer even though no congestion occurs. As a solution to this problem, we suggest not to set FNE for the first and third data packets. TCP SYN packets should still be FNE marked to indicate flow start and prevent end-systems from by-passing the re-insertion of a RE marking by starting a new flow when congestions occurs.

Second, if the marking starts close to the end of one burst, the re-insertion of positive markings cannot be sent any more. If this event occurs several times, the balance at the end of a flow gets more and more negative, as in Fig. 5 for the drop-tail-like RED parametrization. The problem is particularly relevant when the bottleneck marks a large number of packets, e.g., in Slow-Start overshoot. Such a permanent negative balance would cause a re-ECN dropper to penalize a flow even if the sender honestly declares the congestion.

The re-ECN balance also depends on the AQM parameterization, which is still a research challenge [16]. For ECN the total number of marking is mostly irrelevant, as the CWND is reduced at most once per RTT. In contrast, within the re-ECN framework, every marking may cost credits in a policer. Accordingly, different amounts of marks can affect the share of resources obtained by a flow.

5.3   re-ECN Accuracy with Inelastic Cross Traffic

We also investigated a setup where CBR cross traffic competes with a TCP connection. This scenario studies the impact of unresponsive cross traffic on

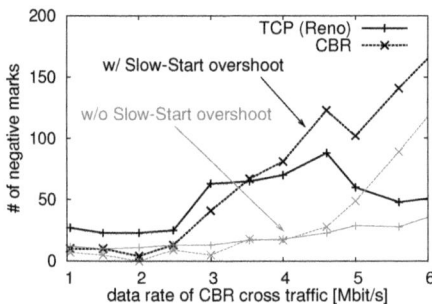 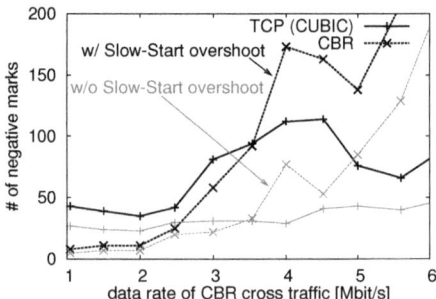

Fig. 6. Number of negative marks with CBR traffic and a TCP Reno sender

Fig. 7. Number of negative marks with CBR traffic and a TCP CUBIC sender

congestion exposure. It is assumed that the CBR traffic also honestly declares its congestion. As congestion exposure should motivate using an appropriate congestion control, one could assume that the CBR traffic will get more re-ECN marks. However, Fig. 6 and 7 show that this is not necessarily true. In order to explore this effect the CBR data rate has been varied in several simulation runs with one minute simulation time each. The results show that CBR traffic can get less markings than a TCP sender when its data rate is small. Only if the CBR data rate is larger than the mean TCP data rate, the CBR traffic requires more re-ECN credits than a TCP connection. This means that re-ECN will not necessarily achieve an enforcement of congestion control.

Furthermore, the higher the data rate of the CBR traffic, the larger is the amount of markings, since the TCP flow then more frequently exceeds the available bandwidth by probing. The gray lines in the diagram show the re-ECN balance values of the same simulation runs but without the markings caused by the Slow-Start overshoot. Again, the number of markings seen by the CBR traffic increases strongly for high CBR rates, as the TCP connection reduces its sending rate more quickly after a congestion event. Thus, even without the Slow-Start effect, the CBR traffic gets a larger share of the overall markings. Apart from that, Figures 6 and 7 also show that, as to be expected, a more aggressive congestion control such as CUBIC results in more re-ECN markings.

5.4   re-ECN Characteristics in a Realistic Internet Traffic Scenario

Finally, we studied a scenario with real Internet traces. Our setup is characterized by a mean load of 45 % in one hour simulation time. Fig. 8 plots the number of negative marks from the bottleneck router for every marked flow in response direction. As most flows in the replayed traces are short, only very few of them actually get marks. As to be expected, the total number of markings is larger for longer flows, simply because they send more packets. However, for a given flow length, the actual number varies over more than one order of magnitude. Because of the statistical nature, one cannot easily estimate the aggressiveness of a user based on the information of one flow only.

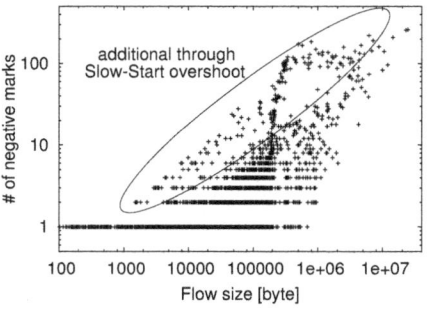

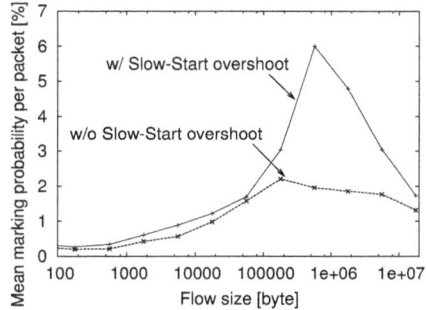

Fig. 8. Number of negative marks over flow size

Fig. 9. Mean probability for a negative marking per data packet over flow size

But there is a further dependency on the flow size: Figure 9 shows the mean number of markings divided by the total number of data packets, as a function of the flow length. Ideally, one would expect about the same marking probability for every packet, but this is not true. This effect is again caused by Slow-Start overshooting. A higher per-packet marking probability can be observed for flows that are just long enough to overshoot. Accordingly, fair capacity sharing through policing might not be achieved, as those flows consume proportionally more of the available congestion volume.

We also analyzed the re-ECN balance at the end of each connection. Only 85.5 % of the flows complete with a re-ECN value of 3 or 2, i.e., they are indeed balanced. 10.7 % of the flows have a larger end balance with a maximum value of +2385, and for 3.51 % the final value is smaller, with a minimum value of -53. These numbers reveal again that non-persistent flows can be permanently unbalanced, e.g. by sending sporadically one to three (FNE marked) packets.

6   Conclusion and Outlook

re-ECN is a proposed TCP/IP extension that exposes the expected congestion on a network path. In this paper we present an own implementation of the re-ECN protocol in the Linux TCP/IP network stack and evaluation results of different simulation scenarios. As re-ECN is just a signaling protocol, we demonstrated that it correctly signals congestion. However, our results show that the information depends on many factors, such as the flow size, the RTT, or AQM parameters. As soon as network components react to re-ECN information, this could imply disadvantages for certain data transports, e.g., with a certain length. We conclude that it is crucial to take such effects into account when interpreting the re-ECN information. In particular, when using the re-ECN information for congestion policing, the absolute number of markings is relevant. Market competition should prevent network providers from overstating congestion and the control mechanism enforced by the dropper should restrain end-systems from understating. Both need further research.

In order to show the advantage of congestion exposure, more research is needed in various subtopics. This includes the design of the proposed network elements (dropper, policer), as well as the usage of the exposed information in the network, e. g., for multi-path routing. Another open issue is the interaction of re-ECN with applications if those would be responsible for realizing an appropriate congestion control, e. g., by adapting the data rate of a video stream or by delaying a file sharing transfer. All these design decisions depend on the investigated characteristics of the re-ECN protocol. The re-ECN protocol also leaves open how a receiver could be made accountable for the congestion caused by a download from a server. Finally, the overall economic implications of congestion exposure for accounting require further studies, e. g., concerning the usage across different network domains and corresponding business models.

References

[1] Allman, M., Paxson, V., Blanton, E.: TCP Congestion Control. RFC 5681, IETF (September 2009)
[2] Ha, S., Rhee, I., Xu, L.: CUBIC: A new TCP-friendly high-speed TCP variant. ACM SIGOPS Operating System Review 42(5), 64–74 (2008)
[3] Bastian, C., Klieber, T., Livingood, J., Mills, J., Woundy, R.: Comcast's Protocol-Agnostic Congestion Management System. Internet draft, IETF (2009)
[4] Briscoe, B., Jacquet, A., Moncaster, T., Smith, A.: Re-ECN: Adding Accountability for Causing Congestion to TCP/IP. Internet draft, IETF (September 2009)
[5] Floyd, S.: TCP and explicit congestion notification. ACM SIGCOMM Computer Communication Review 24(5), 8–23 (1994)
[6] Briscoe, B., Jacquet, A., Di Cairano-Gilfedder, C., Salvatori, A., Soppera, A., Koyabe, M.: Policing Congestion Response in an Internetwork using Re-feedback. ACM SIGCOMM Computer Communication Review 35(4), 277–288 (2005)
[7] Kostopoulos, A. (ed.): D10 - Initial evaluation of social and commercial control progress. Trilogy, EU project FP7 (2009)
[8] Moncaster, T., Krug, L., Menth, M., Araujo, J., Woundy, R.: The Need for Congestion Exposure in the Internet. Internet draft, IETF (2009)
[9] Jansen, S., McGregor, A.: Simulation with Real World Network Stacks. In: Proc. Winter Simulation Conference, pp. 2454–2463 (2005)
[10] Floyd, S., Jacobson, V.: Random Early Detection gateways for Congestion Avoidance. IEEE/ACM Transactions on Networking, 397–413 (August 1993)
[11] Shalunov, S.: Low Extra Delay Background Transport (LEDBAT). Internet draft, work in progress, IETF (March 2009)
[12] Mathis, M.: Relentless Congestion Control. In: Proc. PFLDNeT (2009)
[13] IKR, University of Stuttgart: IKR Simulation and Emulation Library (December 2009), http://www.ikr.uni-stuttgart.de/Content/IKRSimLib/
[14] Andrew, L., Marcondes, C., Floyd, S., Dunn, L., Guillier, R., Gang, W., Eggert, L., Ha, S., Rhee, I.: Towards a common TCP evaluation suite. In: Proc. PFLDnet (2008)
[15] Website: WAN in Lab – Traffic Traces for TCP Evaluation, http://wil.cs.caltech.edu/suit/TrafficTraces.php
[16] Chen, W., Yang, S.-H.: The mechanism of adapting RED parameters to TCP traffic. Computer Communications 32, 1525–1530 (2009)

Estimating AS Relationships for Application-Layer Traffic Optimization

Hirochika Asai and Hiroshi Esaki

The University of Tokyo, Japan
panda@hongo.wide.ad.jp, hiroshi@wide.ad.jp

Abstract. The relationships among autonomous systems (ASes) on the Internet are categorized into two major types: *transit* and *peering*. We propose a method for quantifying AS' network size called *magnitude* by recursively analyzing the AS adjacency matrix converted from a spanning subgraph of the AS-level Internet topology. We estimate the relationships of inter-AS links by comparing differences in magnitude of two neighboring ASes, while showing differences in the magnitude, representing AS relationships appropriately through three evaluations. We also discuss the applicability of this method to AS relationships-aware application-layer traffic optimization.

1 Introduction

The Internet consists of thousands of autonomous systems (ASes) operated by distinct administrative domains such as Internet service providers (ISPs), companies and universities. There are commercial relationships between interconnected ASes, and the relationships are categorized into two major types [1]: *transit* and *peering*. *Transit* relationships are also called provider-customer relationships, and customer ASes purchase Internet access from their transit providers by paying some amount of money. On the contrary, *peering* relationships are equal relationships between interconnected ASes, and traffic exchanged between peering ASes is free of charge. Therefore, transit traffic exchanged with provider ASes costs more for customer ASes compared to that exchanged with customer ASes or traffic exchanged over peering links from the economical viewpoint. Note that we refer to a transit link from a customer AS to a provider AS and a link with opposite orientation as customer-to-provider (c2p) link and provider-to-customer (p2c) link, respectively. We also refer to a peering link as peer-to-peer (p2p) link.

Researches and discussions regarding application-layer traffic optimization have been conducted [2,3]. We propose a path selection method that takes into account the types of AS relationships in content delivery networks utilizing peer-to-peer technologies [4]. We show that the proposed method has reduced high-cost transit traffic for residential ISPs, which provide their network to consumers hosting content delivery network peers, by assigning link cost onto inter-AS links and avoiding selecting high-cost paths. In the proposed method, we have used the types of the relationships to assign cost into inter-AS links. However, there

B. Stiller, T. Hoßfeld, and G.D. Stamoulis (Eds.): ETM 2010, LNCS 6236, pp. 51–63, 2010.

still exists a problem that most commercial ISPs do not want to disclose their relationships because the interconnections are established by their commercial contracts.

Several AS relationships inference algorithms [5,6,7,8] have been proposed. These algorithms infer the relationships by analyzing AS paths in Border Gateway Protocol (BGP) routing tables according to the valley-free path model [9]. However, these algorithms cannot infer the relationships of these invisible links though there are lots of invisible inter-AS links in the set of AS paths extracted from the BGP routing tables because the set of AS paths produce spanning subgraphs (i.e., parts) of the Internet topology and the number of ASes which provide their BGP routing tables to public are limited. Here, we refer to links which are not contained in the set of AS paths in publicly available BGP routing tables as *invisible* inter-AS links. On the other hand, applications on the Internet possibly utilize paths containing invisible inter-AS links as well for their communications because the routing tables of ASes which provide their network to these applications are usually different from the publicly available BGP routing tables. Consequently, it is essential for AS relationships-aware application-layer traffic optimization to estimate the relationships of these invisible links as well as visible links. We note that invisible inter-AS links in an AS path which an application utilizes for its communication can be found by the application with a network management tool (e.g., "traceroute" tool), and the existing algorithms cannot infer the relationships from the found AS path due to lack of AS paths.

In this paper, we propose a method for quantifying the AS' network size, which we call *magnitude*, by recursively analyzing the AS adjacency matrix approximated from a measured spanning subgraph of the AS-level Internet topology according to inter-AS connectivities and a traffic flow model. We show the differences in magnitude appropriately represent AS relationships through three evaluations.

2   AS Relationships Estimation

Relationships between interconnected ASes are characterized by the exchanged traffic volume and the network size [10,11,12]. The traffic volume exchanged over transit links is highly asymmetric, and the traffic volume from a transit provider to the customer is generally larger than that from the customer to the provider. On the other hand, the traffic volume exchanged among peering ASes is nearly symmetric. From the viewpoint of the network size, transit providers are larger than their customers and peering ASes are nearly equal in size as well as exchanged traffic volume. Fig. 1 shows the AS relationships representation by

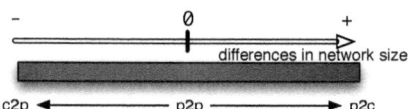

Fig. 1. AS relationships representation by differences in the network size

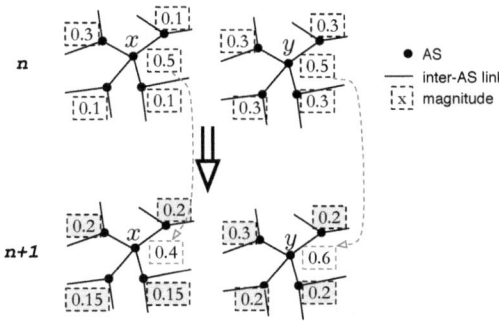

Fig. 2. The concept of recursive definition of magnitude

differences in the network size. Links with positive and those with negative values are considered p2c ones and c2p ones, respectively, and links with values around 0 are considered p2p ones. For example, degree, the number of interconnected ASes, can be used as one of the indicators which represent AS' network size [13]. It is said that differences in degree numerically represent the relationships, and consequently, it has been used for AS relationships inference algorithms based on the path analysis [5,6,7,8].

We propose a method for quantifying AS' network size called *magnitude* which represents AS relationships better than degree. The magnitude is computed recursively by taking into account the magnitude of neighboring ASes to improve the representation of AS' magnitude. Fig. 2 shows the concept of recursive definition of magnitude. In this figure, AS x and AS y have equal magnitude (0.5) where the recursion level is n, but the magnitude of their neighbors is different. This difference makes the magnitude of AS x and AS y different where the recursion level is $n + 1$. Since the neighbors of AS y are larger than those of AS x, AS y become larger than AS x by taking account the magnitude of their neighbors. We then propose a method for estimating the relationships from the quantified magnitude.

2.1   AS Magnitude and Inter-AS Traffic Flow Model

We use a measured spanning subgraph of the AS-level Internet topology for the AS magnitude quantification. Let a graph $G_I = (V_I, E_I)$ be the whole AS-level Internet topology, i.e., a set V_I of ASes contains all ASes on the Internet and a set E_I of inter-AS links contains all inter-AS links though some of them may be invisible from a measurement. We can measure subgraphs of the whole Internet AS graph G_I from AS paths in BGP routing tables. A measured subgraph is generally a quasi spanning subgraph[1] of the AS-level Internet topology. Here,

---

[1] A *quasi* spanning subgraph denotes the subgraph containing *almost all* vertices (ASes) even though it does not contain some edges (inter-AS links). The measured AS paths constitute a quasi spanning subgraph due to the existence of default route configuration and so on; i.e., there exist a few invisible ASes as well as some invisible inter-AS links.

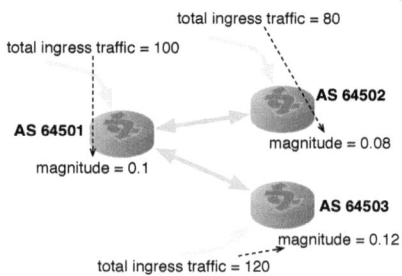

Fig. 3. Definition of AS magnitude

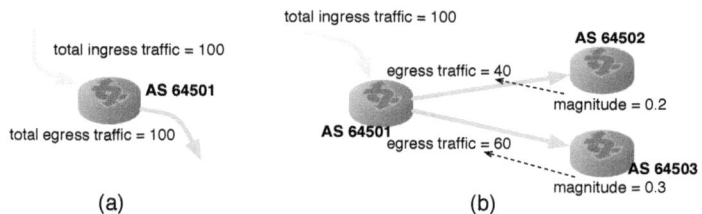

Fig. 4. Traffic flow assumptions

we define a graph $G_S = (V_S, E_S)$ as the measured spanning subgraph. Since the graph G_S is a quasi spanning subgraph, the set V_S of ASes is a subset of the set V_I (i.e., $V_S \subseteq V_I$) and the set E_S of inter-AS links is a subset of the set E_I (i.e., $E_S \subseteq E_I$). A complementary set $\bar{E}_S$ which is the set of invisible inter-AS links are represented by an equation of the form: $\bar{E}_S = E_S \backslash E_I$.

From the measured spanning subgraph G_S, we compute AS' magnitude according to the traffic flow model. In our AS magnitude quantification method, we define that the magnitude is proportional to the total ingress traffic to the AS at the steady state in the traffic flow model as shown in Fig. 3. In another word, the magnitude represents the traffic density at the steady state in the traffic flow model. Let $t_{v_i v_j}$ be traffic from AS v_i to AS v_j, the magnitude of AS v_i is defined by Equation (1). A symbol ρ_{v_i} denotes the magnitude of an AS v_i, subject to equations: $\sum_{v_k \in V_I} \rho_{v_k}^2 = 1$ and $\rho_{v_k \in \bar{V}_S} = 0$.

$$\rho_{v_i} := C \sum_{v_k \in \mathrm{nbr}(v_i)} t_{v_k v_i} \tag{1}$$

$$\text{s.t. } C = const., \sum_{v_k \in V_S} \rho_{v_k}^2 = 1$$

Here, the function $\mathrm{nbr}(v_i)$ returns a set of neighbor ASes of AS v_i, provided that the links between AS v_i and the neighbors are in the set E_S.

We then introduce the simple traffic flow model to compute the magnitude as shown in Fig. 4. We describe the assumptions in the model, as follows:

(a) The total amount of ingress traffic to AS v_i is equal to the egress traffic of AS v_i: $\sum_{v_k \in \mathrm{nbr}(v_i)} t_{v_k v_i} = \sum_{v_k \in \mathrm{nbr}(v_i)} t_{v_i v_k}$.

(b) The amount of egress traffic from AS v_i to AS v_j is proportional to the magnitude of AS v_j (ρ_{v_j}): $t_{v_i v_j} = \frac{\rho_{v_j}}{\sum_{v_k \in \mathrm{nbr}(v_i)} \rho_{v_k}} \sum_{v_k \in \mathrm{nbr}(v_i)} t_{v_i v_k}$.

Since the magnitude is used in these assumptions and it is computed from these assumptions, the magnitude is computed recursively. We compute the steady state of ingress/egress traffic in these assumptions with fixed values of magnitude (e.g., $\boldsymbol{\rho} = [1, \cdots, 1]^t$ for initial case), and then we recursively redetermine the magnitude from the traffic distribution at the steady state.

2.2   AS Magnitude Computation

The steady-state of traffic according to the traffic flow model described in the previous subsection is solved by eigenvalue analysis of a traffic transition matrix. We first define a weighted AS adjacency matrix ^nA by the equation: $^nA := \left(^na_{v_i v_j} \right)$, where $v_i, v_j \in V_S$. Here, the left superscript $^n\bullet$ ($n \geq 0, n \in \mathbb{Z}$) denotes the recursion level. This matrix is extracted from the measured spanning subgraph G_S. Each diagonal element of the matrix ^nA is 0, and other elements are defined by Equation (2a) for initial case ($n = 0$), and by Equation (2b) for other cases.

(i) $n = 0$

$$^na_{v_i v_j} = \begin{cases} 1 & : \text{if AS } v_i \text{ and AS } v_j \text{ are adjacent} \\ 0 & : \text{otherwise} \end{cases} \tag{2a}$$

(ii) $n \geq 1$ ($n \in \mathbb{Z}$)

$$^na_{v_i v_j} = \begin{cases} {}^{(n-1)}\rho_{v_j} & : \text{if AS } v_i \text{ and AS } v_j \text{ are adjacent} \\ 0 & : \text{otherwise} \end{cases} \tag{2b}$$

The matrix ^nA where $n \geq 1$ is defined recursively from the vector of magnitude $^{(n-1)}\rho$. The matrix ^nA where $n \geq 1$ is also represented by the equation: $^nA = I\ ^{(n-1)}\rho\ ^0A$, where I denotes the identity matrix.

Next, we equalize the ingress and egress traffic on each AS by considering that $^na_{v_i v_j}$ represents the egress traffic from AS v_i to AS v_j. We define a traffic transition matrix ^nT by Equation (3).

$$^nT := \left(\frac{^na_{v_i v_j}}{\sum_{v_k} {}^na_{v_i v_k}} \right) \tag{3}$$

We note that the traffic transition matrix is represented by a form of the stochastic matrix. Finally, we compute the steady state of traffic by eigenvalue analysis of the traffic transition matrix. The steady state is determined by calculating the left eigenvector of ^nT corresponding to the maximum eigenvalue. We define this left eigenvector as the vector of magnitude: $^n\boldsymbol{\rho} = [^n\rho_{v_1}, \cdots, {}^n\rho_{v_m}]^t$ (s.t. $\|^n\boldsymbol{\rho}\| = 1$, $m = \|V_S\|$). Here, we note that the magnitude of an AS where $n = 0$ results in a value of the AS' degree multiplied by a constant, though we omit the proof.

2.3 AS Relationships Estimation

From the quantified magnitude, we estimate the relationships of inter-AS links. We define the difference in logarithmic magnitude for an inter-AS link e_x from an AS v_i to an AS v_j as the magnitude distance $^n\delta_{e_x}$. The magnitude distance $^n\delta_{e_x}$ is defined by Equation (4).

$$^n\delta_{e_x} := \log_{10} {}^n\rho_{v_i} - \log_{10} {}^n\rho_{v_j} \tag{4}$$
$$\text{s.t. } e_x = (v_i, v_j),\ e_x \in E_I,\ v_i, v_j \in V_I$$

The distribution (e.g., the minimum and maximum values) of magnitude distances is different for each recursion level n. We can compare the magnitude distances for the same recursion level, but we cannot do it for different recursion levels. To normalize the distribution of magnitude distances to uniform distribution, we define the ranked magnitude distance $^n\delta'_{e_x}$ for a inter-AS link e_x from the magnitude distances. Let a set $^n\delta_S$ be a vector of the magnitude distances for $\forall e \in E_S$, the ranked magnitude distances are defined by Equation (5).

$$^n\delta'_{e_x} := 2\frac{\text{rank-of}(^n\delta_{e_x}) - 1}{\|E_S\| - 1} - 1 \tag{5}$$
$$\text{s.t. } e_x \in E_S$$

Here, the function rank-of returns the rank of the magnitude distance $^n\delta_{e_x}$, sorting magnitude distances in the elements of the vector $^n\delta_S$ in ascending order; i.e., the returned value should be distributed uniformly in the range $[1, \|E_S\|]$. We note that the ranked magnitude distances are hardly applied to application-layer traffic optimization because the ranked magnitude distances are defined only for visible inter-AS links (i.e., $\forall e_x \in E_S$).

The magnitude distance $^n\delta_{e_x}$ numerically represents AS relationships as shown in Fig. 1. For example, if the absolute value $|^n\delta_{e_x}|$ is around 0, two neighboring ASes are nearly symmetric and the inter-AS link e_x is estimated as p2p. Since the magnitude distances numerically represent the relationships, we can infer the type of the relationships from these magnitude distances by setting a threshold in Equation (6).

$$\begin{cases} ^n\delta > {}^n\tau & \to \text{p2c} \\ ^n\delta < -{}^n\tau & \to \text{c2p} \qquad \text{s.t. } {}^n\tau \geq 0 \ (^n\tau\text{: threshold}) \\ -{}^n\tau \leq {}^n\delta \leq {}^n\tau & \to \text{p2p} \end{cases} \tag{6}$$

We note that the magnitude distances can be directly used for AS relationships-aware application-layer traffic optimization without inferring the types of the relationships by Equation (6) as well because they numerically represent the relationships, despite the fact that the inferred types of the relationships can be helpful to give some guidelines to applications.

3   Evaluation

We make three evaluations on the proposed AS relationships estimation method. In the first evaluation, we evaluate the accuracy of the inference of types of the relationships, which are inferred by Equation (6). We show that the types of the relationships are inferred appropriately by the magnitude distances without analyzing AS paths. We also show that the recursive computation of magnitude improves the accuracy of peering inference. In the second evaluation, we show the characteristics of the magnitude distances among well-known tier-1 ISPs. Since the relationships between any two tier-1 ISPs are considered peering, we show these peering links are characterized better by recursive computation of magnitude. In the third evaluation, we show the proposed method can estimate the invisible links by counting the number of the paths which follow the valley-free path model.

3.1   Datasets

We use two types of datasets for the evaluation; 1) *CAIDA's dataset* and 2) *RIB datasets*. CAIDA's dataset defines inter-AS links and the relationships, and RIB datasets define AS paths. CAIDA's dataset is used for both the magnitude computation (i.e., as a quasi spanning subgraph) and the verification (i.e., as a correct AS relationships dataset). RIB datasets are used only for the verification. We describe these datasets below.

Table 1. The number of inter-AS links and the proportion by type of relationships

type of relationships	#links	proportion
sibling (s2s)	219	0.302%
peering (p2p)	6142	8.47%
transit (p2c/c2p)	66181	91.2%

We employ "The CAIDA AS relationships dataset (10/08/2009) [14]" as a quasi spanning subgraph for the magnitude computation and a correct AS relationships dataset for the verification. The relationships in this dataset are inferred by the algorithm [7,8]. In this paper, we call this *CAIDA's dataset*. This dataset contains 32281 ASes and 72542 inter-AS links. We write up the number of inter-AS links and the proportion by type of relationships in Table 1.

We also use Routing Information Base (RIB) datasets (archives: 01/08/2009–05/08/2009) from "Route Views Project [15]" and "RIPE NCC Projects Routing Information Service [16]". We call these *RIB datasets*. We extract AS paths from these datasets, excluding the paths which include private AS numbers, four-octet AS numbers and *AS23456*[2]. We summarize the measurement points, the data

---

[2] Four-octet AS numbers can be translated into *AS23456* when BGP routers do not support four-octet AS numbers. Since we do not analyze BGP options in this paper, we also exclude four-octet AS numbers to identify ASes.

Table 2. Measurement points, the data sources, the number of unique AS paths and the number of unique inter-AS links

measurement point	abbr.	source	#unique paths	#unique links
a) Oregon IX	oregon-ix	RV	1,641,700	69,246
b) Equinix Ashburn	eqix	RV	257,630	57,726
c) ISC (PAIX)	isc	RV	433,861	60,641
d) LINX	linx	RV	784,053	65,774
e) DIXIE (WIDE)	wide	RV	208,542	51,328
f) RIPE NCC, Amsterdam	rrc00	RIS	641,324	64,151
g) Otemachi, Japan (JPIX)	rrc06	RIS	96,951	45,040
h) Stockholm, Sweden (NETNOD)	rrc07	RIS	242,386	56,563
i) Milan, Italy (MIX)	rrc10	RIS	291,297	56,241

"RV" and "RIS" stand for "Route Views Archive Project [15]" and "RIPE NCC Projects Routing Information Service [16]", respectively.

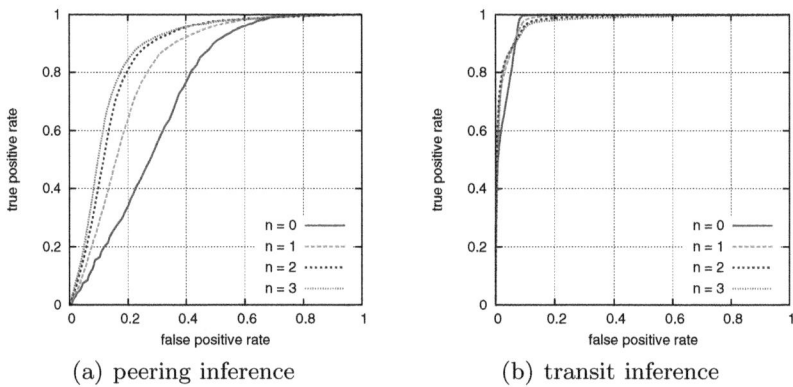

(a) peering inference   (b) transit inference

Fig. 5. ROC curve on inferring peering and transit relationships where $n \in \{0, 1, 2, 3\}$

sources, the number of unique AS paths and the number of unique inter-AS links in Table 2. We use AS paths in these datasets for the verification.

3.2   Evaluation 1: Accuracy of AS Relationships Inference

In this evaluation, we use CAIDA's dataset for both the magnitude computation and the verification. We compute the magnitude for all ASes in the dataset and magnitude distances for all inter-AS links in the dataset. We infer the relationships from the magnitude distances by Equation (6) with sliding the threshold $^{n}\tau$, and verify inferred relationships by those relationships defined in CAIDA's dataset.

We draw a Receiver Operating Characteristic (ROC) curve on inferring peering and transit relationships in Fig. 5; we plot the false positive rate and the true positive rate at x-axis and y-axis, respectively, with sliding the threshold.

Table 3. Well-known tier-1 ISPs

AS no. Name	AS no. Name
7018 AT&T	3549 Global Crossing (GBLX)
3356 Level 3 Communications (L3)	2914 NTT Communications (Verio)
209 Qwest	1239 Sprint
6453 Tata Communications	701 Verizon Business
3561 Savvis	1299 TeliaSonera
6461 AboveNet	2828 XO Communications

It is commonly said that the area under the curve (AUC) represents the accuracy of the inference because points of lower false positive rate and higher true positive rate increases the AUC. This figure shows that the magnitude distances represent both transit and peering relationships appropriately, and the recursive computation improves the accuracy of peering inference. By comparing the AUC on inferring peering, the values of AUC where $n = \{0, 1, 2, 3\}$ are 0.720, 0.814, 0.856 and 0.871, respectively, i.e., the recursive computation improves the accuracy of peering inference. By comparing the AUC on inferring transit, the values of AUC where $n = \{0, 1, 2, 3\}$ are 0.977, 0.982, 0.980 and 0.974, respectively, i.e., the recursive computation does not change the accuracy of transit inference much. From these results, the proposed method with the recursive computation represents the network size and the relationships better than the method without recursive computation (i.e., degree-based one). We note again that the magnitude where $n = 0$ is degree multiplied by a constant.

3.3 Evaluation 2: Characteristics of Ranked Magnitude Distances among Tier-1 ISPs

We showed the recursive computation improves the accuracy of peering inference in the previous subsection. In that evaluation, we assumed that the types of AS relationships defined in CAIDA's dataset are correct, but the relationships in CAIDA's dataset may include inaccurate inferences. In this evaluation, we do not use the relationships defined in CAIDA's dataset to eliminate the influence of inaccurate inferences in CAIDA's dataset while we use CAIDA's dataset for both the magnitude computation. Instead, we evaluate the links among well-known tier-1 ISPs. The links between any two tier-1 ISPs are considered peering. We list the well-known tier-1 ISPs in Table 3.

We show the characteristics of AS relationships of the inter-AS links among well-known tier-1 ISPs in Fig. 6. Each line represents a link between two neighboring tier-1 ISPs. The relationships between any two tier-1 ISPs are considered peering. Hence, Fig. 6 shows the characteristics of magnitude differences of peering links. This figure shows that the recursive computation of magnitude decreases the absolute value of ranked magnitude distance $|{}^{n}\delta'|$. This means the recursive computation of magnitude improves the accuracy of peering estimation. the maximum values among the ranked magnitude distances of the links between tier-1 ISPs are 0.178 where $n = 0$ and 0.0732 where $n = 5$.

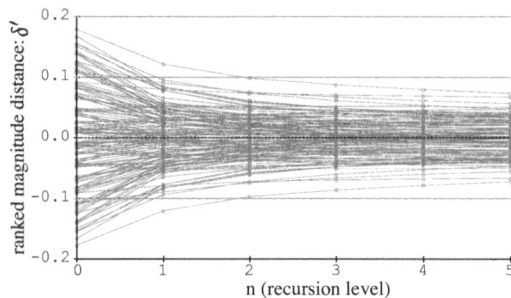

Fig. 6. Characteristics of AS relationships of inter-AS links among well-known tier-1 ISPs

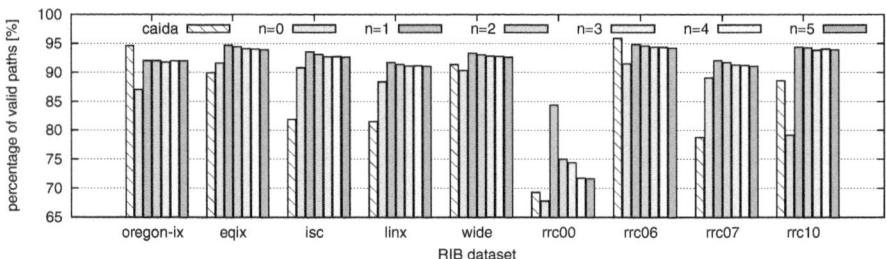

Fig. 7. Percentage of valid paths, which follow the valley-free path model

3.4   Evaluation 3: AS Relationships Estimation for Invisible Links

In this evaluation, we use RIB datasets for the verification based on the valley-free path model [9]. We annotate AS relationships to all the inter-AS links in RIB datasets by Equation (6) with the threshold: $^{n}\tau = 0$. To show the advantage of the proposed method, we annotate AS relationships to them by the relationships defined in CAIDA's dataset as well. We then count valid paths (i.e., paths following valley-free path model). When we consider that paths between any two ASes follow the valley-free path model, the number of valid paths represents the accuracy of AS relationships inference.

We show the percentage of valid paths for each RIB dataset in Fig. 7. A legend *caida* denotes the paths are annotated by the relationships defined in CAIDA's dataset, and the other legends $n = \{0, \cdots, 5\}$ denote the paths are annotated by the inferred relationships from the magnitude distances. We note that larger values, in this figure, represent higher accuracy on inferring AS relationships. Excluding the paths annotated by the relationships defined in CAIDA's dataset, the percentage of valid paths annotated by the magnitude distances where $n = 1$ is highest for every RIB dataset. This means the orientations of transit links are represented the best by the magnitude distances where $n = 1$. Additionally, including the paths annotated by the relationships defined in CAIDA's dataset, the percentage of valid paths annotated by the magnitude distances where $n = 1$ is highest for all of the datasets except oregon-ix and rrc06. Though the relationships in CAIDA's dataset are inferred so as to maximize the number of valid paths, the percentage is not highest. This is because there are some links which are not contained in CAIDA's dataset, but the magnitude distances can be computed from the quantified magnitude if the edge ASes are contained in the magnitude computation procedure.

4   Discussion

Inter-AS traffic flow model: We introduce a simple inter-AS traffic flow model in Section 2.1 to compute the magnitude, but the actual inter-AS traffic flow is not so simple. For example, we assume the total ingress traffic volume is equal to the total egress traffic volume but the total ingress traffic volume is generally larger than the total egress traffic volume at residential ISPs. To justify the traffic flow model, we discuss the meaning of the model for the recursion level $n = 0$. For the recursion level $n = 0$, the traffic transition matrix ^{0}T becomes a stochastic matrix; i.e., this model implies random walk-like transition of traffic. We have described that the magnitude of an AS where $n = 0$ results in a value of the AS' degree multiplied by a constant. For the recursion levels $n \geq 1$, the traffic transition matrix ^{n}T is weighted by the magnitude of neighbors, and the weighting procedure matches the hierarchical routing on the Internet; i.e., traffic tends to go to transit providers because the transit providers (larger ASes) relay traffic to other ASes. Therefore, the traffic flow model is justified by the definition of degree and the hierarchical routing. There is no doubt that the traffic flow model and the weighting procedure can be modified to improve the proposed method. We will work on this modification in future.

The recursion level: The recursion level means the hop count to which the method takes into account the network size for the magnitude computation. For example, $n = 1$ means the method takes into account the network size of neighbors, and $n = 2$ means the method takes into account the network size of neighbors and that of two-hop neighbors. From Evaluation 1 and 2, we show that the accuracy of peering inference is improved by increasing the recursion level n. On the other hand, from Evaluation 3, the orientations of transit links are represented the best where the recursion level is $n = 1$. These results show that the orientations of transit links are represented by the network size from the local viewpoint (i.e., at most one-hop neighbors' size), and the peering links are represented by the network size from the global viewpoint.

AS relationships-aware application-layer traffic optimization: We have described that the proposed method is applicable to AS relationships-aware application-layer traffic optimization. Suppose, for instance, there are two content mirror servers s_1 and s_2, and a client c, and the path from s_1 to c and that from s_2 to c are $\{s_1 \to \text{p2p} \to c\}$ and $\{s_2 \to \text{c2p} \to \text{p2c} \to c\}$, respectively. To reduce high-cost transit traffic, the client should select the server s_1. The relationships of each inter-AS link between s_1 and c, and s_2 and c are required to be inferred to enable AS relationships-aware server selection. The magnitude distances or the inferred relationships by Equation (6) can be used for it. As described in Section 2.3, the ranked magnitude distances are hardly applied to application-layer traffic optimization. The magnitude distances are easily applied to applications by using values of these distances as metric directly. When we use the inferred relationships, the threshold should be tuned for each application; e.g., some applications permit false positive and the others do not.

Paid peer consideration: On the Internet, there are so-called *paid peering* relationships, which are intermediate relationships between peering and transit. The proposed method can quantify the network size well, and the relationships are characterized by the magnitude distances. Hence, we do consider the possibility of estimating the paid peer relationships as well as the applicability applicability of the proposed method for estimating these complex relationships in future.

5 Related Work

Gao [5] has proposed an algorithm to infer AS relationships. The author has shown that the relationships can be inferred by comparing the number of neighbors (i.e., degree) between two neighboring ASes, analyzing the AS paths in BGP routing tables based on the valley-free path model [9]. Battista et al. [6] improved Gao's algorithm. They mapped this problem into weighted MAX2SAT (maximum-2-satisfiability) problem to compute the orientation of transit links. However, on the real Internet, there are lots of invisible inter-AS links. Therefore, it is difficult to apply these AS relationships inference algorithms based on path analysis to AS relationships-aware application-layer traffic optimization because applications often utilize links which relationships are not annotated by these algorithms from the spanning subgraphs.

6 Summary

We proposed a method for quantifying the AS' network size called *magnitude* by recursively analyzing the AS adjacency matrix which is approximated from a measured spanning subgraph of the AS-level Internet topology. We showed that the differences in magnitude appropriately represent AS relationships by three evaluations. We also showed that the recursive computation of magnitude improved the accuracy of peering inference. The contributions of this paper are followings: 1) The proposed method can estimate AS relationships of any inter-AS links, and the estimated relationships are applicable to AS relationships-aware application-layer traffic optimization. 2) The proposed method uses AS adjacency information which is more highly available information than AS paths which have been commonly used in the previous works.

We will apply the estimated magnitude distances to AS relationships-aware application-layer traffic optimization, and design an architecture to utilize these distances as a traffic control metric.

Acknowledgment

We especially thank Kensuke Fukuda and Yosuke Himura for their valuable advice on the AS graph analysis. We also thank Burkhard Stiller for shepherding us to improve this paper.

References

1. Shakkottai, S., Srikant, R.: Economics of network pricing with multiple ISPs. IEEE/ACM Trans. Netw. 14(6), 1233–1245 (2006)
2. Xie, H., Yang, Y.R., Krishnamurthy, A., Liu, Y.G., Silberschatz, A.: P4P: provider portal for applications. In: SIGCOMM 2008: ACM SIGCOMM 2008 Conference on Data Communication, pp. 351–362. ACM, New York (2008)
3. The Internet Engineering Task Force (IETF): Application-layer traffic optimization (alto), http://datatracker.ietf.org/wg/alto/charter/
4. Asai, H., Esaki, H.: A content delivery path control architecture for the multi-domain P2P CDN system. In: Internet Conference 2008 (October 2008) (in Japanese)
5. Gao, L.: On inferring autonomous system relationships in the Internet. IEEE/ACM Transactions on Networking 9(6), 733–745 (2001)
6. Battista, G.D., Erlebach, T., Hall, A., Patrignani, M., Pizzonia, M., Schank, T.: Computing the types of the relationships between autonomous systems. IEEE/ACM Trans. Netw. 15(2), 267–280 (2007)
7. Dimitropoulos, X., Krioukov, D., Huffaker, B., Claffy, K.C., Riley, G.: Inferring AS relationships: Dead end or lively beginning? In: Nikoletseas, S.E. (ed.) WEA 2005. LNCS, vol. 3503, pp. 113–125. Springer, Heidelberg (2005)
8. Dimitropoulos, X., Krioukov, D., Fomenkov, M., Huffaker, B., Hyun, Y., Claffy, K.C., Riley, G.: As relationships: inference and validation. SIGCOMM Comput. Commun. Rev. 37(1), 29–40 (2007)
9. Erlebach, T., Hall, A., Panconesi, A., Vukadinovi, D.: Cuts and disjoint paths in the valley-free path model of Internet BGP routing. In: López-Ortiz, A., Hamel, A.M. (eds.) CAAN 2004. LNCS, vol. 3405, pp. 49–62. Springer, Heidelberg (2005)
10. Norton, W.B.: Internet service providers and peering (2000)
11. Weiss, M.B., Shin, S.J.: Internet interconnection economic model and its analysis: Peering and settlement. Netnomics 6(1), 43–57 (2004)
12. Huang, C., Li, J., Ross, K.W.: Can internet video-on-demand be profitable? In: SIGCOMM 2007: Proceedings of the 2007 Conference on Applications, Technologies, Architectures, and Protocols for Computer Communications, pp. 133–144. ACM, New York (2007)
13. Tangmunarunkit, H., Doyle, J., Govindan, R., Willinger, W., Jamin, S., Shenker, S.: Does AS size determine degree in AS topology? SIGCOMM Comput. Commun. Rev. 31(5), 7–8 (2001)
14. Cooperative Association for Internet Data Analysis: The CAIDA AS Relationships Dataset (10/08/2009), http://www.caida.org/data/active/as-relationships/
15. University of Oregon: Route Views Project, http://www.routeviews.org/
16. Réseaux IP Européens Network Coordination Centre: Routing Information Service (RIS), http://www.ripe.net/projects/ris/

Mobile Internet in Stereo: An End-to-End Scenario

Henna Warma[1], Tapio Levä[1], Lars Eggert[2], Heikki Hämmäinen[1], and Jukka Manner[1]

[1] Aalto University, Department of Communications and Networking,
P.O. Box 13000, 00076 Aalto, Finland
{henna.warma,tapio.leva,heikki.hammainen,jukka.manner}@tkk.fi
[2] Nokia Research Center, P.O. Box 403, 00045 Nokia Group, Finland
lars.eggert@nokia.com

Abstract. Multipath TCP (MPTCP) is an extension to regular TCP that exploits the idea of resource pooling by transmitting the data of a single TCP connection simultaneously across multiple Internet paths between two end hosts. Operating system vendors are traditionally in the key position to facilitate the deployment of new functionality, such as MPTCP, to user devices, but their motivation to invest in such enhancements is not self-evident. A scenario in which one party has the capability to deploy software changes on both the mobile devices and on the content servers helps to understand the potential first-mover advantages associated with the deployment of a new IETF standard protocol. In this study, we have built a quantitative techno-economic model to estimate the implementation costs for a content provider selling application downloads who is also able to implement MPTCP on both ends and compare them against the revenue MPTCP generates. The results suggest that even a relatively small increase in the number of downloads could make the business case profitable within five years.

Keywords: Multipath TCP, deployment, content provider, techno-economic modeling.

1 Introduction

One intriguing change to the Internet protocol suite that has gained significant standardization traction recently is Multipath TCP (MPTCP). It is an extension to standard TCP that supports the simultaneous use of multiple network paths between two end hosts involved in a TCP connection. Thus, MPTCP is a manifestation of resource pooling principle that improves utilization of Internet resources by allowing separate resources to behave as if they were a single aggregate resource [1]. MPTCP uses a coordinated congestion control scheme [2] to balance transmission rates across the paths it transmits over in order to guarantee fairness to standard TCP while increasing connection throughput and resilience. The standardization of MPTCP is currently ongoing in the Internet Engineering Task Force (IETF) [3].

The deployment of any change to an Internet protocol is a complex process; the deployment of a new protocol even more so. Although a new protocol may bring substantial verified benefits and may have a good technical design, these factors alone

B. Stiller, T. Hoßfeld, and G.D. Stamoulis (Eds.): ETM 2010, LNCS 6236, pp. 64–75, 2010.

are often not sufficient to guarantee successful deployment of the protocol. The deployment of such protocols depends on the incentives for and the behavior of other stakeholders in the broader Internet market. When a new protocol is available for end users, they may never have to come to a conscious decision to start using it but the new protocol functionality can be enabled for them, *e.g.*, through software updates.

This paper investigates MPTCP deployment and adoption using a specific scenario. Of particular interest is whether MPTCP offers a profitable business case for a client-server type of stakeholder controlling the operating systems on both mobile devices and on the content servers. Multiple examples of this scenario exist. Nokia, Apple, Microsoft and Google all sell portable devices to consumers and also provide service portals for users to access and download content, such as music, movies and applications.

A techno-economic model, *i.e.*, a net present value (NPV) analysis, is used to quantify the business case by comparing the costs of implementing MPTCP against the revenue of the additional business that MPTCP may generate. The model considers a mobile application store and uses generic values that are based on real-life references to the Nokia Ovi Store and the Apple App Store. The results indicate that even a small increase in downloads due to MPTCP turns the investment profitable. Exploiting scale advantages is possible, because CAPEX dominates OPEX and in CAPEX, software implementation costs are significantly higher than hardware costs.

2 Background

This section introduces the reader to the existing work related to MPTCP deployment and techno-economic modeling.

2.1 MPTCP Deployment

Because MPTCP is a strict end-to-end protocol, an implementation needs only affect the end points. Ref. [4] presents the overall three-step deployment process using different technology and Internet standards adoption theories. First, operating system (OS) vendors need to make the protocol implementation available. Second, the protocol needs to be deployed on mobile devices and data centers. End users can either directly install MPTCP or an installation may happen indirectly, *e.g.,* through automatic software updates and new device purchases. Third, at least one of the two ends needs to multihome, *i.e.*, to connect to the Internet through multiple paths, which may require contracting with multiple Internet service providers (ISP).

Although the involvement of ISPs is not mandatory for either the implementation or the deployment of MPTCP, Ref. [5] describes how ISPs may support the adoption of MPTCP in different deployment scenarios. The model of this paper does not discuss how end users have contracted with ISPs to become multihomed, but the role of ISPs in the middle of MPTCP communication is discussed in Section 6.

In addition to the direct implementation by end hosts, MPTCP can also be deployed via proxies [6][7]. MPTCP proxies can be deployed upstream of end hosts that only implement standard TCP; they terminate inbound MPTCP connections and relay the communication over standard TCP in a way that is transparent to both ends. The result is that an MPTCP host believes it is communicating with another MPTCP host, while the other peer believes the communication uses standard TCP. Proxies can also be deployed in front of both end hosts [8], in which case a communication session across the network can be "upgraded" to MPTCP without modifying the end hosts at all.

2.2 Techno-economic Modeling

The techno-economic modeling exploits future forecasting, technology design and investment theories in evaluating the economic feasibility of a new technology. Usually, a NPV analysis is used, because it takes the time value of money into account and therefore lends itself well to estimating future cash flows. Different authors have developed spreadsheet-based tools specifically intended for the telecommunications industry. Katsianis *et al.* [9] use techno-economic modeling for estimating the evolution of mobile network operators, whereas Smura *et al.* [10] use it to model the financial position of mobile virtual network operators (MVNO). Because techno-economic modeling includes forecasting and usually many of the assumptions are uncertain, it should always be followed by a sensitivity analysis.

3 Implementation Scenario

Because MPTCP is an end-to-end protocol, both clients and servers require MPTCP support before the protocol can be used. This section identifies the implementation requirements for both ends and discusses how a content provider can deploy MPTCP.

Two high-level approaches exist for content providers to implement MPTCP. The first option is to implement MPTCP through a multihomed proxy that is installed at the border of the data center. MPTCP is then used between the mobile end user and the proxy, which forwards the communication to the actual content servers over a regular TCP connection. The second option is to update the load balancers at the edge of the data center to support MPTCP, deploy MPTCP on the servers in the data center, and also multihome those servers. The load balancers will need to be updated, because otherwise they may forward the individual subflows of one MPTCP connection to separate servers in the data center, therefore rendering MPTCP ineffective.

This paper assumes that MPTCP is deployed on the data center side using proxies. Fig. 1 illustrates the technical architecture of this scenario. The dashed line shows the multipath connection. A gradual rollout of MPTCP proxies is possible but the full benefits of MPTCP are realized only after all content transfers use MPTCP. An advantage of the proxy implementation is that the internal service infrastructure of the data center can remain unchanged and that the load balancers, which forward user requests to different content servers, do not require any changes.

Although Fig. 1 depicts the MPTCP proxy as having multiple network connections, multihoming is not a mandatory requirement at the content server side, *if* the mobile user is multihomed (which this paper assumes). However, data center multihoming increases the number of possible connection subflows and many data centers are multihomed for fault tolerance already.

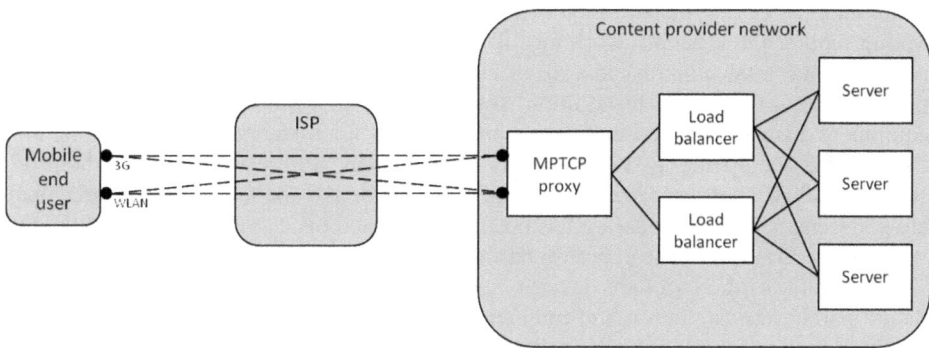

Fig. 1. Technical architecture of the scenario

A mobile end user can obtain MPTCP support either by installing MPTCP to her end host and acquiring multiple access connections to become multihomed, or through an MPTCP proxy provided by her ISP [5]. The scenario where an end user is multihomed is more promising, because the involvement of ISPs is not needed and MPTCP functionality can be implemented solely on the mobile device. Therefore, this paper does not consider proxying on the client side.

The basic functionality of MPTCP is implemented as an extension to TCP in the transport layer of the operating system kernel. In order for MPTCP to remain compatible with existing applications, an MPTCP implementation needs to operate effectively when accessed through the standard TCP API.

Although most existing hardware platforms already support multiple network access technologies, some operating systems do not expose the availability of multiple accesses above the IP layer. This prevents higher-layer protocols and applications from using multiple paths. Such operating systems require additional modifications to enable simultaneous use of multiple access connections.

4   Modeling Provider Costs and Benefits

Evaluating the profitability of an investment compares the costs against the revenue and other benefits. This section discusses the costs and benefits for the content provider deploying MPTCP. The basic assumptions of the techno-economic model are also presented.

4.1   Costs

Implementing the MPTCP functionality in operating systems causes the largest expenses to content providers. Content servers typically run different or at least differently optimized operating systems than mobile devices. MPTCP needs to be implemented for each operating system in use, which is costly, but distributing the software to devices incurs only low costs.

A software development project consists of three main steps: feature specification, implementation and testing. The implementation phase is the most time consuming phase and requires more workforce than the other two. In the model, the implementation phase takes three months and the other phases two months. Software specification and testing require three people working full time, whereas implementation requires eight people. If the total monthly cost of an employee is 10 000 €, completing these three phases costs 440 000 €. In addition, installation of the software as well as testing equipment incurs some costs for the provider, so an assumption for implementing MPTCP for one operating system is 500 000 €. The model presumes that on the proxy side, the MPTCP software is implemented once and on the end user side, for three operating systems. After the software has been implemented once, costs of developing other versions are lower, because expertise has been acquired during the first implementation.

In addition to the software development costs, deployment of MPTCP on proxies incurs purchase, installation and maintenance costs. The level of these costs depends on the complexity of the content server infrastructure and the number of required proxies. Additional network connections may be required to satisfy the multihoming requirement, but this paper presumes that server farms are already sufficiently multihomed, so that this cost component can be neglected.

The number of required proxies depends on the capacity of each proxy. Modern servers appear to be capable of handling at least 20 000 simultaneous TCP sessions (*e.g.*, [11]). Because an MPTCP connection consists of multiple TCP subflows, the capacity of an MPTCP server is assumed to be 10 000 simultaneous MPTCP sessions. Interviewed experts in the field suggested that a proxy with this capacity costs approximately 5 000 €. The installation costs are included in the price. In order to avoid a single point of failure in the content service, the required number of proxies is doubled in the model.

The operational expenditures of a server include maintenance work such as fixing fault situations and updating the software on occasion. The energy consumption cannot be neglected and it presumably dominates OPEX compared to the maintenance costs. Ref. [12] suggests that the annual energy costs of a small server are approximately 1 500 €. The server type used in the model has larger energy consumption. The annual maintenance costs are initially 30 % and the energy costs are 50 % of the purchase price of a server. The model assumes that maintenance costs increase in relation to the growth rate of MPTCP users in the content provider's network, but energy costs remain stable during the study period, because the servers do not support power control. Table 1 summarizes the cost assumptions of the model.

Table 1. Cost assumptions of the model

Item	Assumption
CAPEX	
Implementing MPTCP (datacenter side)	500 000 €
Implementing MPTCP (end user side)	1 200 000 €
Proxy purchasing	5 000 €
OPEX	
Initial maintenance costs of a proxy	1 500 €
Energy costs of a proxy (per year)	2 500 €

4.2 Benefits

Analyzing the benefits for the content provider is not as straightforward as analyzing the costs. This is because end users may not be willing to pay directly for MPTCP functionality on their mobile device. Therefore, a content provider should not expect any revenue to come from simply providing such an update to the end users. However, content users experience increased throughput and resilience when using MPTCP for accessing content. These benefits translate into shorter download times and more robust connectivity, especially when mobile. Thus, the fundamental assumption of the model is that deploying MPTCP will increase demand, *i.e.*, the number of content accesses.

On the other hand, using two radio connections simultaneously may also increase battery consumption and thus decrease the number of downloads. However, in light of [13], the effect of battery consumption on the number of downloads should not be significantly negative. Until MPTCP has been implemented and tested in real deployments, the benefits of MPTCP cannot be verified.

Although MPTCP will likely increase the quality of experience (QoE), the question is whether this will lead to additional content downloads. Ref. [14] indicates that an increase in bandwidth has a positive effect on time spent online. This does not mean that end users would spend all this additional time downloading content, but the assumption is that a fraction of time is used for that purpose.

However, more user activity does not necessarily translate into more profits. End users may be charged according to different pricing models that also affect the usage and consequently provider revenue. Three different pricing models exist in the content delivery business. Flat rate pricing encourages a user to access content as much as he desires but the increasing usage increases only costs – not revenue. In transaction-based pricing, a content provider gets additional revenue from each additional download, but users consider more carefully what content they access. In the third case, a content provider offers free services to customers, and gets revenue from somewhere else, typically through advertisements. Although advertisers are charged according to the service usage, the revenue cannot be easily attributed to downloads.

The model considers only application downloads, because they are charged per transaction and their revenue can therefore be easily verified. Another argument for choosing applications is that users most probably download them straight to mobile devices. Music, on the other hand, is presumably often first downloaded to a regular computer and from there transferred to a mobile device.

Browsing Apple's [15] and Nokia's [16] offerings reveals that only a small fraction of applications are being charged for. Therefore, a parameter describing the amount of chargeable downloads is included in the model. Ref. [17] suggests 25 % of downloaded applications from Apple's App Store require payment. This percentage seems quite substantial and MPTCP will unlikely increase the number of chargeable downloads by the same ratio. A user's buying decision is affected by many factors and the impact of QoE is seen to be a matter of secondary importance. Therefore, the proportion of chargeable downloads of all increased downloads is 1 % in the model.

Ref. [18] summarizes the average application prices in different application stores and reveals that the average price of all chargeable applications is 3.62 € for Apple

Table 2. Assumptions concerning the revenue

Parameter	Value
Percentage of downloads chargeable	1 %
Average price of a chargeable application	3.50 €
Provider's profit of an application price	30 %

and 3.47 € for the Nokia. An average price of 3.50 € has been used in the model. The entire price of a sold item does not go to the content provider. Ref. [17] suggests that the application developers get a 70 % cut of revenue while the provider gets 30 %. Table 2 summarizes the assumptions concerning the revenue.

Deploying MPTCP can benefit a content provider also in ways that are not directly related to an increased usage of content. The increased throughput and resilience of MPTCP connections may increase device sales, especially if MPTCP becomes more widely adopted. Additionally, if MPTCP can optimize the battery consumption required for Internet connectivity, the device manufacturer can use the conserved energy to improve the quality of user experience, *e.g.,* through extended battery life or new attractive features. However, these benefits are not considered in the model.

5 Techno-economic Model

This section proposes a simplified, quantitative, techno-economic model that integrates the different incremental costs and benefits to estimate the profitability of MPTCP for a content provider. MPTCP is only used for content downloads; all other communication between the end user and the content provider, such as signing in to the service, uses regular TCP. The basic assumptions of the model have been presented in Section 4, but this section explains how the traffic volumes in the content provider's network have been calculated. The model is supposed to be general and therefore does not represent the business case of any specific content provider.

The model assumes a study period of five years and an interest rate of 10 %, both of which are often used in techno-economic modeling. Both the network dimensioning and revenue calculations are based on the estimated number of downloads per day from the content service. Daily download figures of Nokia's Ovi Store (1.5 million) [19] and Apple's App Store (6.6 million) [20] from the first half of 2010 have been used as references. The initial number of downloads per day is an average of these two (4 million).

The growth of application stores has been accelerating heavily. For example, [21] shows that the number of downloads from Apple's App Store has been increasing by approximately 500 % per year. An explanation of such a heavy growth rate is that sales of iPhone devices have also been growing considerably. Once the device sales have been saturated the growth rate of application downloads will be much smaller. The model in question presumes an annual linear growth of 25 %.

Like any other new innovation, MPTCP is adopted gradually. Ref. [22] shows that the diffusion of new features on mobile handsets follows an "s-shaped" curve. The authors in [22] have chosen Gompertz model [23] to analyze the diffusion patterns. The same model is adapted here to correspond to assumed progress of MPTCP diffusion. The suggested model has two parameters that define the shape of the curve. The displacement parameter *b* has been chosen to be 0.8 and the growth rate parameter *c* is set to 5. After five years, the penetration of MPTCP increases up to 90 %. The diffusion curve models relatively fast penetration, which is reasonable because as an end-to-end provider the OS vendor can significantly increase the penetration rate of MPTCP.

The required number of proxy servers can be calculated when the maximum number of simultaneous MPTCP sessions per second and the capacity of a proxy are known. The maximum number of simultaneous sessions has been deduced from the average number of downloads per second using a dimensioning rule that the average traffic load is 40 % of the maximum load. Based on different statistics of traffic amounts on different Internet links, 40 % appears to be a valid estimate. The traffic is assumed to spread equally over all geographical areas. The average number of simultaneous sessions is calculated using Equation 1.

$$Avg\ number\ of\ sessions/s = \frac{Number\ of\ downloads\ per\ day * Avg\ duration\ of\ a\ session}{Seconds\ per\ day} \quad (1)$$

The average duration of a session can be further deduced from the average size of an application and the average throughput of a session. Ref. [15] is used to estimate the average size of an application (10 MB) and measurements with a mobile phone supporting 3G and WLAN to define the average throughput of a session (4000 Kb/s).

5.1 Results and Sensitivity Analysis

The model reveals that in order for the NPV of the investment to be zero, MPTCP should increase downloads by approximately 3.5 % per MPTCP user. If a content user currently downloads one chargeable application per month, he spends an average of 42 € per year on downloading applications. Consequently, MPTCP should make the user spend less than an additional 2 € on applications per year.

The number of daily application downloads reaches 12.2 million during the study period. This traffic volume can be managed with two proxies and thus both CAPEX and OPEX of proxies remain small. In addition, the model shows that software implementation costs dominate CAPEX instead of hardware costs.

To have a perception how MPTCP affects a provider's revenue, the increasing demand of applications is illustrated in Fig. 2. The figure shows how the provider's NPV varies as a function of MPTCP impact on downloads. The solid line shows the base scenario with the values presented above.

Because several assumptions of the model come with significant uncertainty, a sensitivity analysis is performed. The analysis reveals that the most critical parameter in the model is the percentage of the chargeable items among additional downloads. An intriguing fact is that this is also the most uncertain parameter in the model. To illustrate the effect of this parameter, Fig. 2 also shows how NPV is affected when the percentage of chargeable items varies.

Five years may be a relatively long period for an investment like MPTCP, because it is only one software update among many others. With a payback time of 3 years, the base scenario approximately results in a 12 % increase in downloads due to MPTCP. If we assume the same behavior from the user as earlier, each MPTCP user should spend around 5 € more on applications per year.

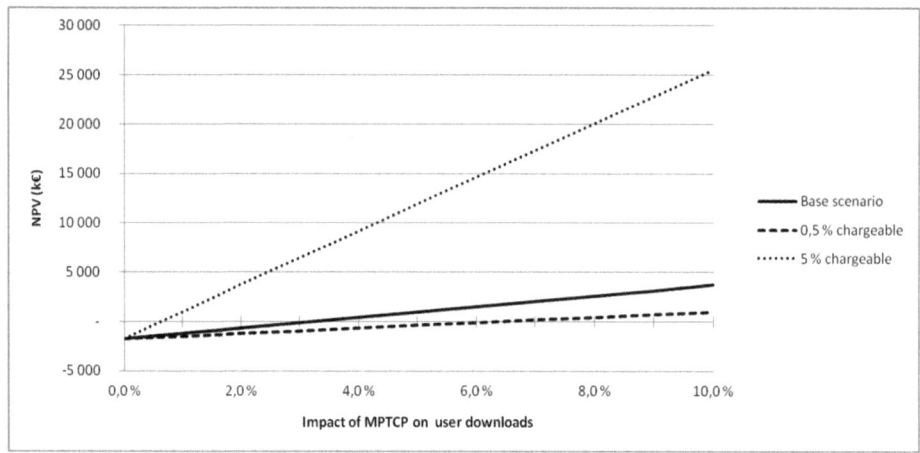

Fig. 2. NPV of the investment as a function of MPTCP impact on downloads

The sensitivity analysis suggests that the provider's revenue may be much bigger than shown in Fig. 2. The reason is that the increased demand of applications affects the revenue more strongly than the costs of the provider. In addition, the other sources of revenue have not been considered in the model, such as increased sales of other types of content or mobile devices, which can further increase the provider's revenue.

6   Discussion

Although the presented techno-economic model suggests that MPTCP is a profitable investment for content providers, it relies heavily on the assumption that increased bandwidth increases content downloads. Even if this would be the case with free content, the assumption can be questioned with paid content. The selection of content is naturally the most important factor but also the price, ease of paying and general usability of the service affect the purchase decision probably much more than download speed.

The paper also assumes that end users are always interested in maximizing throughput. This is reasonable with application downloads but, for example, for video streaming, the bandwidth only needs to exceed the video peak rate, and faster download rates are not very useful to any stakeholder. Thus, depending on the service, users may actually be more interested in, for example, minimizing Internet connectivity costs or energy consumption. This paper assumes flat rate pricing in all access connections, but if, for example, a 3G connection is usage- or block-priced, end users will favor flat rate priced WLAN connections over 3G, if they are available.

The access pricing model also affects the ISPs' stake in MPTCP. Using multiple access paths simultaneously can reduce the load per single path, which can either be a desired or an undesired effect. If both access connections are purchased from the same ISP, the ISP will benefit if the traffic moves from a more congested access link to a less congested one. The pricing model dictates the impact of load: with flat rate pricing, the operator's profit increases when traffic decreases, and with usage-based pricing, more traffic increases revenues. Because ISPs are in the middle of an end-to-end MPTCP communication, they have also the possibility of interrupting or blocking MPTCP traffic, if it seems harmful for their business.

Finally, it should be noted that the transport layer is not the only option for implementing benefits of MPTCP. If different access connections have highly asymmetrical throughput, the usefulness of using multiple connections at the same time can be questioned. A solution that would provide prioritizing of access connections and would enable smooth handover can offer similar benefits without the need to change the networking stack. Furthermore, new radio access technologies will increase the available bandwidth but the bottleneck can also be located elsewhere than in the access network. The availability of multiple paths affects the usefulness of MPTCP, because multiple access paths are ineffective if all share the bottleneck link.

MPTCP-like functionality could also be implemented directly by applications. A multi-server HTTP mechanism [24] increases the efficiency of network resource utilization and Adobe's proprietary Real-Time Media Flow Protocol (RTMFP) [25] provides a real-time, peer-to-peer bulk transport protocol over UDP. RTMFP is provided as part of Adobe's Flash plugin for web browsers, which currently makes it unsuitable for general-purpose applications, but significantly eases deployment and adoption. MPTCP is a more comprehensive solution that requires operating system support but is available to all applications on the host. This tradeoff between generality and deployability poses an interesting question in which layer the benefits of resource pooling principle should actually be trialed and implemented.

7 Conclusion

This paper has analyzed the profitability of an MPTCP deployment scenario where a single provider controls both the mobile devices and content servers, and implements and adopts MPTCP. The model suggests that even a relatively small increase in download numbers due to MPTCP could make the business case profitable for such a provider. Although the results of the model are very uncertain, this is a positive signal for MPTCP. All the example providers of the scenario are corporations with wide customer bases and getting these players "on board" to implement MPTCP is hence highly important.

Although the model is simple, it gives a good understanding of different cost and revenue components incurred by MPTCP and serves as a good base for further research. The model should be extended to also cover other types of content and pricing schemes. We chose to currently limit the scenario to only a single provider, because the linkage between costs and revenue can be easily investigated. Taking into account multiple content providers as well as other deployment scenarios are worth of further study.

A fundamental question regarding the current and potential future models is whether MPTCP really can improve the quality of experience for Internet services. Although the benefits can be verified, it is unclear whether better service quality will lead to an increase in service use. Actual measurements as well as studies on user behavior would be needed to answer these questions.

Acknowledgments. This paper was partly funded by *Trilogy*, a research project supported by the European Commission under its Seventh Framework Program (INFSOICT-216372). The research is also linked to the European COST IS 0605 / Econ@Tel project.

References

1. Wischik, D., Handley, M., Bagnulo Braun, M.: The Resource Pooling Principle. ACM SIGCOMM CCR 38(5), 47–52 (2008)
2. Wischik, D., Raiciu, C., Handley, M.: Balancing Resource Pooling and Equipoise in Multipath Transport. Submitted to ACM SIGCOMM (2010)
3. Ford, A., Raiciu, C., Barre, S., Iyengar, J.: Architectural Guidelines for Multipath TCP development. IETF Internet-Draft (work in progress), draft-ietf-mptcp-architecture-00 (2010)
4. Kostopoulos, A., Warma, H., Levä, T., Heinrich, B., Ford, A., Eggert, L.: Towards Multipath TCP Adoption: Challenges and Opportunities. In: 6th Euro-NF Conference on Next Generation Internet, Paris, France (2010)
5. Levä, T., Warma, H., Ford, A., Kostopoulos, A., Heinrich, B., Widera, R., Eardley, P.: Business Aspects of Multipath TCP Adoption. Future Internet Assembly Book (2010)
6. Balakrishnan, H., Seshan, S., Amir, E., Katz, R.H.: Improving TCP/IP Performance over Wireless Networks. In: 1st Annual International Conference on Mobile Computing and Networking, Berkeley, USA (1995)
7. Henderson, T.R., Katz, R.H.: Transport protocols for Internet-compatible satellite networks. IEEE Journal on Selected Areas in Communications 17(2), 326–344 (1999)
8. Border, J., Kojo, M., Griner, J., Montenegro, G., Shelby, Z.: Performance Enhancing Proxies Intended to Mitigate Link-Related Degradations. IETF RFC 3135 (2001)
9. Katsianis, D., Welling, I., Ylönen, M., Varoutas, D., Sphicopoulos, T., Elnegaard, N.K., Olsen, B.T., Burdry, L.: The financial perspective of the mobile networks in Europe. IEEE Personal Communications 8(6), 58–64 (2001)
10. Smura, T., Kiiski, A., Hämmäinen, H.: Virtual Operators in the mobile industry: a techno-economic analysis. NETNOMICS 8(1-2), 25–48 (2008)
11. Dodownload.com: Faststream IQ Proxy server, http://www.dodownload.com/servers/other+servers/fastream+iq+reverse+proxy.html (accessed 2010-04-27)
12. Electronics Cooling: In The data center, power and cooling costs more than the it equipment it supports, http://www.electronics-cooling.com/2007/02/in-the-data-center-power-and-cooling-costs-more-than-the-it-equipment-it-supports/ (accessed 2010-06-10)
13. Nurminen, J.K.: Parallel Connections and their Effect on the Battery Consumption of a mobile phone. In: 7th IEEE Consumer Communications & Networking Conference, Las Vegas, Nevada (2010)
14. Anderson, B.: The Social Impact of Broadband Household Internet Access. Information, Communication & Society 11(1), 5–24 (2008)
15. Apple App Store, http://www.apple.com/iphone/apps-for-iphone/ (accessed 2010-04-14)
16. Nokia Ovi Store, http://store.ovi.com/ (accessed 2010-04-14)
17. GigaOm: Apple App Store Economy, http://gigaom.com/2010/01/12/the-apple-app-store-economy/ (accessed 2010-04-02)

18. Distimo report: Apple App Store, BlackBerry App World, Google Android Market Nokia Ovi Store & Windows Marketplace for Mobile, `http://www.distimo.com/uploads/reports/Distimo%20Report%20-%20December%202009.pdf` (accessed 2010-06-11)
19. Freak, S.: Ovi Store Statistics: 22 Application downloads per second, `http://www.symbian-freak.com/news/010/03/ovi_store_stats.htm` (accessed 2010-04-12)
20. MediaMemo: Apple's Apps Flying Off the Virtual Shelves: 6.6 Million downloads per day, `http://mediamemo.allthingsd.com/20090928/apples-apps-flying-off-the-virtual-shelves-6-6-million-downloads-per-day/` (accessed 2010-04-12)
21. Macrumors.com: Apple announces 3 billion App Store Downloads, `http://www.macrumors.com/2010/01/05/apple-announces-3-billion-app-store-downloads/` (accessed 2010-04-26)
22. Kivi, A., Smura, T., Töyli, J.: Diffusion of Mobile Handset Features in Finland. In: 8th International Conference on Mobile Business, Dalian, China (2009)
23. Gompertz, B.: On the nature of the function expressive of the law of human mortality, and on a new mode of determining the value of life contingencies. Philosophical Transaction Series I 115, 513–583 (1825)
24. Ford, A., Handley, M.: HTTP Extensions for Simultaneous Download from Multiple Mirrors. IETF Internet-Draft (work in progress), draft-ford-http-multi-server-00 (2009)
25. Kaufman, M.: RTMFP Overview. IETF-77, Anaheim, CA, USA, `http://www.ietf.org/proceedings/10mar/slides/tsvarea-1.pdf` (accessed 2010-04-12)

A Study of Non-neutral Networks with Usage-Based Prices

Eitan Altman[1], Pierre Bernhard[1], Stephane Caron[2], George Kesidis[2,*],

Julio Rojas-Mora[3], and Sulan Wong[4]

[1] INRIA. 2004 Route des Lucioles,
06902 Sophia-Antipolis, France
{eitan.altman,pierre.bernhard}@sophia.inria.fr
[2] CS&E and EE Depts, Pennsylvania State Univ. University Park, PA, 16802
spc15@psu.edu, Kesidis@engr.psu.edu
[3] Fac. of Econ. and Bus. Sci., Univ. of Barcelona,
08034 Barcelona, Spain
jrojasmo7@alumnes.ub.edu
[4] Fac. of Law, Univ. of Coruña. 15071 A Coruña, Spain
swong@udc.es

Abstract. Hahn and Wallsten [1] wrote that network neutrality "usually means that broadband service providers charge consumers only once for Internet access, do not favor one content provider over another, and do not charge content providers for sending information over broadband lines to end users." We study the implications of non-neutral behaviors under a simple model of linear demand-response to *usage-based* prices. We take into account advertising revenues for the content provider and consider both cooperative and non-cooperative scenarios. We show that by adding the option for one provider to determine the amount of side payment from the other provider, not only do the content provider and the internaut suffer, but also the Access Provider's performance degrades.

Keywords: Network neutrality, game theory.

1 Introduction

Network neutrality is an approach to providing network access without unfair discrimination among applications, content or traffic sources. Discrimination occurs when there are two applications, services or content providers that require the same network resources, but one is offered better quality of service (shorter delays, higher transmission capacity, *etc.*) than the other. How to define what

---

* The work by George Kesidis is supported in part by NSF CNS and Cisco Systems URP grants.

B. Stiller, T. Hoßfeld, and G.D. Stamoulis (Eds.): ETM 2010, LNCS 6236, pp. 76–84, 2010.

is "fair" discrimination is still subject to controversy[1]. A preferential treatment of traffic is considered fair as long as the preference is left to the user[2]. Internet Service Providers (ISPs) may have interest in traffic discrimination either for technological or economic purposes. Traffic congestion, especially due to high-volume peer-to-peer traffic, has been a central argument for ISPs against the enforcement of net neutrality principles. However, it seems many ISPs have blocked or throttled such traffic independently of congestion considerations.

ISPs recently claimed that net neutrality acts as a disincentive for capacity expansion of their networks. In [2], the authors studied the validity of this argument and came to the conclusion that, under net neutrality, ISPs invest to reach a social optimal level, while they tend to under/over-invest when neutrality is dropped. In their setting, ISPs stand as winners while content providers (CPs) are left in a worse position, and users who pay the ISPs for preferential treatment are better off while other consumers have a significantly worse service.

In this paper, we focus on violations of the neutrality principles defined in [1] where broadband service providers charge consumers more than "only once" through usage-based pricing, and charge content providers through side payments. Within a simple game-theoretic model, we examine how regulated[3] side payments, in either direction, and demand-dependent advertising revenues affect equilibrium usage-based prices. We also address equilibria in Stackelberg leader-follower dynamics. We finally study the impact of letting one type of provider determine the amount of side payment from the other provider, and show that this results in bad performance to both providers as well as internauts.

The rest of the paper is organized as follows. In Section 2, we describe a basic model and derive Nash equilibria for competitive and collaborative scenarios.

---

[1] The recent decision on Comcast v. the FCC was expected to deal with the subject of "fair" traffic discrimination, as the FCC ordered Comcast to stop interfering with subscribers' traffic generated by peer-to-peer networking applications. The Court of Appeals for the District of Columbia Circuit was asked to review this order by Comcast, arguing not only on the necessity of managing scarce network resources, but also on the non-existent jurisdiction of the FCC over network management practices. The Court decided that the FCC did not have express statutory authority over the subject, neither demonstrated that its action was "reasonably ancillary to the [...] effective performance of its statutorily mandated responsibilities". The FCC was deemed, then, unable to sanction discriminatory practices on Internet's traffic carried out by American ISPs, and the underlying case on the "fairness" of their discriminatory practices was not even discussed.

[2] Nonetheless, users are just one of many actors in the net neutrality debate, which has been enliven throughout the world by several public consultations for new legislations on the subject. The first one was proposed in the USA, the second one was carried out in France and a third one is intended to be presented by the EU during summer 2010. See [5,3,4].

[3] In the European Union, dominating positions in telecommunications markets (such as an ISP imposing side-payments to CPs at a price of his choice) are controlled by the article 14, paragraph 3 of the Directive 2009/140/EC, considering the application of remedies to prevent the leverage of a large market power over a secondary market closely related.

We consider potentially non-neutral side-payments in Section 3 and advertising revenues in Section 4, analyzing in each case how they impact equilibrium revenues. In Section 5, we study leader-follower dynamics. In Section 6 we study the results of allowing one provider to control the amoount of side payments from the other one. We conclude in Section 7 and discuss future work.

2   Basic Model

Our model encompasses three actors, the internauts (users), collectively, a network access provider for the internauts, collectively called ISP1, and a content provider and its ISP, collectively called CP2. The two providers play a game to settle on their (usage-based) prices. The internauts are modeled through their demand response. They are assumed willing to pay a usage-based fee (which can be $0/byte) for service/content that requires both providers.

Denote by $p_i \geq 0$ the usage-based price leveed by provider i (ISP1 being $i = 1$ and CP2 being $i = 2$). We assume that the demand-response of customers, which corresponds to the amount (in bytes) of content/bandwidth they are ready to consume given prices p_1 and p_2, follows a simple linear model:

$$D = D_0 - d(p_1 + p_2). \tag{1}$$

With such a profile, we are dealing with a set of homogeneous users sharing the same response coefficient d to price variations. Parameter D_0 corresponds to the demand under pure *flat-rate* pricing ($p_1 = 0 = p_2$).

Demand should be non-negative, *i.e.*,

$$p_1 + p_2 \leq \frac{D_0}{d} =: p_{\mathsf{max}}.$$

Provider i's usage-based revenue is given by

$$U_i = D p_i. \tag{2}$$

2.1   Competition

Suppose the providers do not cooperate. A Nash Equilibrium Point (NEP) (p_1^*, p_2^*) of this two-player game satisfies:

$$\frac{\partial U_i}{\partial p_i}(p_1^*, p_2^*) = D^* - p_i^* d = 0 \quad \text{for } i = 1, 2,$$

which leads to $p_1^* = p_2^* = D_0/(3d)$. The demand at equilibrium is thus $D^* = D_0/3$ and the revenue of each provider is

$$U_i^* = \frac{D_0^2}{9d}. \tag{3}$$

2.2 Collaboration

Now suppose there is a coalition between ISP1 and CP2. Their overall utility is then $U_{\text{total}} := U_1 + U_2 = Dp$, and an NEP (p_1^*, p_2^*) satisfies

$$\frac{\partial U_{\text{total}}}{\partial p_i}(p_1^*, p_2^*) \;=\; D^* - d(p_1^* + p_2^*) \;=\; 0 \quad \text{for } i = 1, 2,$$

which yields $p^* := p_1^* + p_2^* = D_0/(2d)$. The demand at equilibrium is then $D^* = D_0/2$, greater than in the non-cooperative setting. The overall utility $U_{\text{total}}^* = D_0^2/(4d)$ is also greater than $D_0^2/(4.5d)$ for the competitive case. Assuming both players share this revenue equally (trivially, the Shapley values are $\{1/2, 1/2\}$ in this case), the utility per player becomes

$$U_i^* = \frac{D_0^2}{8d}, \tag{4}$$

which is greater than in the competitive case. So, both players benefit from this coalition.

3 Side-Payments under Competition

Let us suppose now that there are *side payments* between ISP1 and CP2 at (usage-based) price p_s. The revenues of the providers become:

$$U_1 = D\,(p_1 + p_s)\;;\; U_2 = D\,(p_2 - p_s) \tag{5}$$

Note that p_s can be positive (ISP1 charges CP2 for "transit" costs) or negative (CP2 charges ISP1, *e.g.*, for copyright remuneration[4]). It is expected that p_s is *not* a decision variable of the players, since their utilities are monotonic in p_s and the player without control would likely set (usage-priced) demand to zero to avoid negative utility. That is, p_s would normally be *regulated* and we will consider it as a fixed parameter in the following (with $|p_s| \leq p_{\text{max}}$).

First, if $|p_s| \leq \frac{1}{3}p_{\text{max}}$, the equilibrium prices are given by

$$p_1^* = \frac{1}{3}p_{\text{max}} - p_s\,;\, p_2^* = \frac{1}{3}p_{\text{max}} + p_s$$

---

[4] In France, a new law has been proposed recently to allow download of unauthorized copyright content, and in return be charged *proportionally* to the volume of the download [9]. A similar law had been already proposed and rejected five years ago by the opposition in France. It suggested to apply a tax of about five euros on those who wish to be authorized to download copyrighted content. In contrast, the previously proposed laws received the support of the trade union of musicians in France. If these laws were accepted, the service providers would have been requested to collect the tax (that would be paid by the internauts as part of their subscription contract).

but demand $D^* = D_0/3$ and utilities

$$U_i^* = \frac{D_0^2}{9d}$$

are exactly the same as (3) in the competitive setting with no side payment. Therefore, though setting $p_s > 0$ at first seems to favor ISP1 over CP2, it turns out to have no effect on equilibrium revenues for both providers.

Alternatively, if $p_s \geq \frac{1}{3}p_{\mathsf{max}}$, a boundary Nash equilibrium is reached when $p_1^* = 0$ and $p_2^* = \frac{1}{2}(p_{\mathsf{max}} + p_s)$, which means ISP1 does not charge usage-based fees to its consumers. Demand becomes $D^* = \frac{1}{2}(D_0 - dp_s)$, and utilities are

$$U_1^* = \frac{(D_0 - dp_s)dp_s}{2d} \; ; \; U_2^* = \frac{(D_0 - dp_s)^2}{4d}$$

Though $p_1^* = 0$, U_1^* is still strictly positive, with revenues for ISP1 coming from side-payments (and possibly from flat-rate monthly fees as well). Furthermore, $p_s \geq \frac{1}{3}p_{\mathsf{max}} \Leftrightarrow dp_s \geq \frac{1}{2}(D_0 - dp_s)$, which means $U_1^* \geq U_2^*$: in this setting, ISP1's best move is to set his usage-based price to zero (to increase demand), while he is sure to achieve better revenue than CP2 through side-payments.

Finally, if $p_s < -\frac{1}{3}p_{\mathsf{max}}$, the situation is similar to the previous case (with $-p_s$ instead of p_s). So, here $p_2^* = 0$ and $p_1^* = \frac{1}{2}(p_{\mathsf{max}} - p_s)$, leading to $U_2^* \geq U_1^*$.

To remind, herein revenues U_i are assumed usage-based, which means there could also be flat-rate charges in play to generate revenue for either party. Studies of flat-rate compare to usage-based pricing schemes can be found in the literature, see, *e.g.*, [6].

4   Advertising Revenues

We suppose now that CP2 has an additional source of (usage-based) revenue from advertising that amounts to Dp_a. Here p_a is not a decision variable but a fixed parameter.[5]

4.1   Competition

The utilities for ISP1 and CP2 are now:

$$U_1 = [D_0 - d \cdot (p_1 + p_2)](p_1 + p_s) \qquad (6)$$
$$U_2 = [D_0 - d \cdot (p_1 + p_2)](p_2 - p_s + p_a) \qquad (7)$$

Here, the Nash equilibrium prices are:

$$p_1^* = \frac{1}{3}p_{\mathsf{max}} - p_s + \frac{1}{3}p_a \; ; \; p_2^* = \frac{1}{3}p_{\mathsf{max}} + p_s - \frac{2}{3}p_a$$

---

[5] One may see p_a as the result of an independent game between CP2 and his advertising sources, the details of which are out of the scope of this paper.

The cost to users is thus $p^* = \frac{2}{3}p_{\max} - \frac{1}{3}p_a$ while demand is $D^* = \frac{1}{3}(D_0 + dp_a)$. Nash equilibrium utilities are given by

$$U_i^* = \frac{(D_0 + dp_a)^2}{9d} \quad \text{for } i = 1, 2, \tag{8}$$

which generalizes equation (3) and shows how advertising revenue quadratically raises players' utilities.

4.2   Collaboration

The overall income for cooperating providers is

$$U_{\text{total}} = (D_0 - dp)(p + p_a). \tag{9}$$

So, solving the associated NEP equation yields

$$p^* = \frac{p_{\max} - p_a}{2}. \tag{10}$$

The NEP demand is then $D^* = (D_0 + dp_a)/2$, and the total revenue at Nash equilibrium is $U_{\text{total}}^* = (D_0 + dp_a)^2/(4d)$. Assuming this revenue is split equally between the two providers, we get for each provider the equilibrium utility

$$U_i^* = \frac{(D_0 + dp_a)^2}{8d}, \tag{11}$$

which generalizes equation (4). As before, providers and users are better off when they cooperate.

Thus, we see that $p_a > 0$ leads to lower prices, increased demand and more revenue for *both* providers (*i.e.*, including ISP1).

5   Stackelberg Equilibrium

Stackelberg equilibrium corresponds to asymmetric competition in which one competitor is the leader and the other a follower. Actions are no longer taken independently: the leader takes action first, and then the follower reacts.

Though the dynamics of the games are different from the previous study, equations (6) and (7) still hold, with fixed $p_a \geq 0$ and regulated p_s. In the following, we need to assume that

$$p_s \leq \frac{1}{2}p_{\max} + \frac{1}{2}p_a \; ; \; p_a \leq \frac{1}{3}p_{\max} + \frac{1}{4}p_s$$

so that NEPs are reachable with positive prices.

If ISP1 sets p_1, then CP2's optimal move is to set

$$p_2 = \frac{1}{2}(-p_1 + p_{\max} + p_s - p_a).$$

This expression yields $D = \frac{d}{2}(p_{max} - p_1 - p_s + p_a)$ and $U_1 = \frac{d}{2}(p_{max} - p_1 - p_s + p_a)(p_1 + p_s)$. Anticipating CP2's reaction in trying to optimize U_1, the best move for ISP1 is thus to set

$$p_1^* = \frac{1}{2}p_{max} - p_s + \frac{1}{2}p_a \rightarrow p_2^* = \frac{1}{4}p_{max} + p_s - \frac{3}{4}p_a.$$

Therefore, when ISP1 is the leader, at the NEP demand is $D^* = \frac{1}{4}(D_0 + dp_a)$ and utilities are:

$$U_1^* = \frac{1}{8d}(D_0 + dp_a)^2 \; ; \; U_2^* = \frac{1}{16d}(D_0 + dp_a)^2. \tag{12}$$

Suppose now that CP2 is the leader and sets p_2 first. Similarly, we find:

$$p_2^* = \frac{1}{2}p_{max} + p_s - \frac{1}{2}p_a \; ; \; p_1^* = \frac{1}{4}p_{max} - p_s + \frac{1}{4}p_a$$

These values yield the same cost p^* and demand D^* for the internauts at the NEP, while providers' utilities become:

$$U_1^* = \frac{1}{16d}(D_0 + dp_a)^2 \; ; \; U_2^* = \frac{1}{8d}(D_0 + dp_a)^2. \tag{13}$$

Therefore, in either case of leader-follower dynamics, the leader obtains twice the utility of the follower at the NEP (yet, his revenue is not better than in the collaborative case).

6   Further Abandoning Neutrality

Throughout we assumed that the side payments between service and content providers were regulated. If p_s is allowed to be determined unilaterally by the service provider or by the content provider (as part of the game described in Section 3 or 4) then the worst possible performance is obtained at equilibrium. More precisely, the demand at equilibrium is zero, see [8]. The basic reason is that if the demand at equilibrium were not zero then a unilateral deviation of the provider that controls the side payment results in a strict improvement of its utility. (Note that the demand is unchanged as it does not depend on p_s.)

More generally, assume that an ISP is given the authority to control p_s and that its utility can be expressed as $U = f(D) \times (g(p_s) + h)$ where f is any function of the demand (and possibly also of prices other than p_s) g is a monotone strictly increasing function of p_s, and h does not depend on p_s. Then at equilibrium, necessarily $f(D) = 0$, otherwize U can be further increased by the provider by increasing (unilaterally) p_s.

The same phenomenon holds also in case the CP is given full control over p_s.

7   Conclusions and On-Going Work

Using a simple model of linearly diminishing consumer demand as a function of usage-based price, we studied a game between a monopolistic ISP and a

CP under a variety of scenarios including consideration of: non-neutral two-sided transit pricing (either CP2 participating in network costs or ISP1 paying for copyright remuneration), advertising revenue, competition, cooperation and leadership.

In a basic model without side-payments and advertising revenues, both providers achieve the same utility at equilibrium, and all actors are better off when they cooperate (higher demand and providers' utility).

When regulated, usage-based side-payments p_s come into play, the outcome depends on the value of $|p_s|$ compared to the maximum usage-based price $p_{\max}$ consumers can tolerate:

- when $|p_s| \leq \frac{1}{3}p_{\max}$, providers shift their prices to fall back to the demand of the competitive setting with no side-payments;
- when $|p_s| \geq \frac{1}{3}p_{\max}$, the provider receiving side payments sets its usage-based price to zero to increase demand, while it is sure to be better off than his opponent.

When advertising revenues to the CP come into play, they increase the utilities of *both* providers by reducing the overall usage-based price applied to the users. ISP1 and CP2 still share the same utility at equilibrium, and the increase in revenue due to advertising is quadratic.

Under leader-follower dynamics, the leader obtains twice the utility of his follower at equilibrium; yet, he does not achieve a better revenue than in the cooperative scenario.

We finally showed that by adding the option for one provider, say the service provider, to determine side payments from the other provider, not only do the content providers and the internauts suffer, but also the Access Provider's performance degrades.

Acknowledgements

The work by Penn State is supported in part by the National Science Foundation under grant 0916179 and by a Cisco Systems URP gift.

References

1. Hahn, R., Wallsten, S.: The Economics of Net Neutrality. Economists' Voice 3(6), 1–7 (2006)
2. Cheng, K., Bandyopadhyay, S., Gon, H.: The debate on net neutrality: A policy perspective. Information Systems Research (June 2008), SSRN: http://ssrn.com/abstract=959944
3. Consultation publique sur la neutralité du net, du 9 avril au 17 mai (2010), http://www.telecom.gouv.fr/fonds_documentaire/ consultations/10/consneutralitenet.pdf
4. Neelie Kroes Vice President of the European Commission, Address at TEFAF ICT Business Summit, Press release RAPID (April 2010), http://europa.eu/rapid/pressReleasesAction.do?reference=SPEECH/10/87

5. Notice of proposed rulemaking (FCC 09/93) (October 2009–April 2010)
6. Kesidis, G., Das, A., de Veciana, G.: On Flat-Rate and Usage-based Pricing for Tiered Commodity Internet Services. In: Proc. CISS, Princeton (March 2008)
7. Kesidis, G.: Congestion control alternatives for residential broadband access by CMTS. In: Proc. IEEE/IFIP NOMS, Ōsaka, Japan (April 2010)
8. Altman, E., Bernhard, P., Kesidis, G., Rojas-Mora, J., Wong, S.: A study of non-neutral networks. INRIA Research report Number 00481702 (May 2010)
9. Zumkleller, M.: Proposition de loi sur la création d'une licence globale á palier, visant á financer les droits d'auteur dans le cadre d'échanges de contenus audiovisuels sur internet (travaux préparatoires), nro 2476, déposée le 29 avril (2010)

Stability of Alliances between Service Providers

Hélène Le Cadre*

PRISM, Université de Versailles Saint-Quentin,
45, avenue des Etats-Unis, 78035 Versailles, France
`Helene.Le-Cadre@prism.uvsq.fr`

Abstract. Three service providers in competition, try to optimize their quality of service/content level and their service access price. But, they have to deal with uncertainty on the consumers' preferences. To reduce their uncertainty, they have the opportunity to buy information and to build alliances. We determine the Shapley value which is a fair way to allocate the grand coalition's revenue between the service providers. Then, we identify the values of β (consumers' sensitivity coefficient to the quality of service/content) for which allocating the grand coalition's revenue using the Shapley value guarantees the system stability.

Keywords: Alliance; Shapley value; Stability.

1 Introduction

We speak about alliances when firms on a market agree with one another to realize profits which are superior to the *standard* profits that they should receive under competition. The *standard* profits are those obtained at the non cooperative Nash equilibrium where each firm tries to maximize selfishly his utility.

Under competition, firms maximize their profits. But, sometimes, firms realize that coordination might increase their joint profit. Hence, firms on a market have natural incentives to agree together in order to increase their market power and their profit. Competition between firms can then be compared with a prisoner dilemma: a firm chooses his strategy in order to maximize his profit but, does not care about the effect of his decision on the other firms. In an alliance, firms take into account how their own decisions affect the others' profits. There exists of course various forms of alliances: joint price determination, joint quantity setting, geographic allocation, etc.

Cases of collusion (explicit or tacit) have been reported in the telecommunication literature, both in the long distance market and in the mobile sector. In practice, alliances are difficult to build since firms might suspect that their partner would defect and then, adopt tit for tat strategy which, in a short time-scale, does not give incentives to firms for cooperate. Besides, it is difficult to deal with the selfish tendancy of the alliance partners. These latter might for instance, enter a learning race where the firm who learns the quickest wins and

* This work has been funded by the european FP 7 project ETICS (Economics and Technologies for Inter-Carrier Services).

B. Stiller, T. Hoßfeld, and G.D. Stamoulis (Eds.): ETM 2010, LNCS 6236, pp. 85–92, 2010.

then, leaves the alliance. Furthermore, alliances may be highly instable due to changes in customers' demand, political relation and alliance management. de Man et al. propose a robust framework based on a win-win strategy to guarantee the long-term alliance stability in the airline industry [1]. However, reputation phenomenon and guarantee provisioning might give firms incentives to stay in the alliance [2].

One of the main difficulty studied in cooperative game theory [4] is the pie sharing to guarantee that none of the alliance member would have incentives to leave it. Various allocation of the grand coalition's revenue rules exist such as the Shapley value which provides a fair sharing, the nucleolus which provides an allocation that minimizes the dissatisfaction of the players from the allocation that they can receive, the proportional allocation which shares the total revenue depending on the players' intial investment costs, the equal allocation which gives an equal proportion to each player, etc. [4].

The article aims at studying the alliances that might emerge between three service providers in competition, who have to cope with uncertainty on the consumers' preferences. The originality of our approach is to incorporate uncertainty about the consumers' preferences in the game between the operators. Besides, to our knowledge, this is the first article proposing solutions to deal with alliance instability in the telecommunication framework and modelling explicitly information exchange/investment between the operators.

The article is organized as follows. We start by introducing the two level game between the service providers in Section 2. Then, in Section 3, we compute the Shapley value of the cooperative game which enables us to allocate fairly the grand coalition's revenue between the three service providers. Depending on the consumers' sensitivity to the quality of service (QoS)/contents (β), we determine then, whether the Shapley value belongs to the core of the game i.e., guarantees the alliance stability. We conclude in Section 4.

2   Game Model and Information Sharing

Description of the providers' utilities. We consider three service providers: A_1, A_2 and A_3. They can belong to various categories. It can be content providers, Internet access providers, or TV channels, etc. We assume that these three firms are interconnected via the Internet backbone. Besides, they have the opportunity to exchange information. The information can be either bought from other providers or the providers can choose to invest together to gather data.

Let the firm A_i, $i = 1, 2, 3$. His utility is of the form $U_i = p_i n_i - \mathcal{I}(\alpha_i)$ with p_i the access price to the firm i's network/service for the consumers[1], n_i the total number of clients for firm A_i. α_i represents the level of information collected/bought by firm A_i while $\mathcal{I}(.)$ is the cost of information acquisition for the firms.

---

[1] This access price is a flat rate i.e., a fixed price which does not depend on the quantity of traffic really sent by the clients.

The QoS/content level q_i can be seen as a function of firm A_i's information level i.e., $\alpha_i = \vartheta(q_i)$ with $\vartheta : \mathbb{R}_+ \to \mathbb{R}_+$ an invertible function on $\mathbb{R}_+$. Furthermore, we assume that $\mathcal{I} \circ \vartheta$ is convex on $\mathbb{R}_+$. The QoS/content level is tightly related to the information level acquired by the firm. Indeed, if firm A_i manages to extract the most pertinent information about the network topologies or more generally, about the consumers' preferences, he will be able to optimise q_i more easily. Of course the consumers' perception of firm A_i will by turns depend on the firm's information level since, in this case, the firm's information level conditions his QoS/content level. Furthermore, we assume that the firm is limited in his investment level in the QoS/contents by his access network capacity; thus: $q_i \leq q_i^{\max}$, $i = 1, 2, 3$.

In practice, service providers reserve bandwidth on transit providers' networks who route their traffics. These transit operators have approximately identical marginal costs. It is then essential for the service providers to book the optimal bandwidth volume on transit providers' networks using the acquired information to infer consumers' preferences. Indeed, service providers will have to pay for the quantity of data transferred. Besides, it is essential for service providers to reserve the exact quantity of bandwidth; otherwise transit providers punish them either by evening out their traffic if the reservation is too small or by overbooking and deprioritarizing if it is too high. It might then altered the QoS/content level perceived by the consumers.

Game description for firm A_i. The three service providers play simultaneously the two level game described below. Then, depending on their choices, the consumers subscribe to a service or report their decision. The two levels in the game result from the difference of timing between the access price determination and the contract of bandwidth reservation.

(1) A_i sets his QoS/content level q_i
(2) A_i chooses the access price to his service p_i

Description of the consumers' choice model. We suppose that the consumers have the opportunity to choose their service provider or to not subscribe to any service. In order to make their choice, they should compute the opportunity costs [3] associated with each service provider. For firm A_i, the opportunity cost is $c_i = p_i + \beta q_i$ with $-1 \leq \beta \leq 0$ coefficient characterizing consumers' sensitivity to the QoS/content level.

The consumer choice model description requires the introduction of intrinsic utilities [3] for the consumers, associated with each provider, being independent of the opportunity costs. Consumers' intrinsic utility for firm A_i will be denoted $\mathcal{U}_i$. Furthermore, we assume that the consumers have an a priori preference for the firms which is characterized by the following order $\mathcal{U}_1 \leq \mathcal{U}_2 \leq \mathcal{U}_3$. The order is arbitrary and it can be modified. We assume that $0 \leq c_1(.) < c_2(.) < c_3(.) < 1$. This hypothesis guarantees the existence of positive market shares for each of the three firms. An approaching choice model for the consumers has already been introduced in [3].

Then, we introduce the maximum admissible opportunity cost for consumer k, $k = 1, 2, ..., N$: Θ_k, whose realization θ_k is generated according to the uniform density on $[0; 1]$. We have supposed that θ_k was drawn according to a uniform density since the firms have a priori no information on the consumers' preferences. Necessity to introduce a maximum admissible opportunity cost results from the following observation: a consumer will refuse to subscribe to the provider's service or will report his purchase to a later date for the following reasons, either the access price is too high, or the QoS level is not sufficient, or the contents are not enough diversified compared to what he expects. Consumer k's utility for firm A_i, $i = 1, 2, 3$ is $u_{k,i} = \mathcal{U}_i$ if $\theta_k \geq c_i(.)$ and 0 otherwise. This means that if $\theta_k \in [0; c_1(.)[$ consumer k does not choose any offer, if $\theta_k \in [c_1(.); c_2(.)[$, consumer k selects firm A_1, if $\theta_k \in [c_2(.); c_3(.)[$, consumer k chooses firm A_2, finally, if $\theta_k \in [c_3(.); 1]$, the consumer selects firm A_3. By generalization, we can compute the number of clients for each firm. Let n_0 be the total number of consumers who choose to not subscribe to any service, n_i, the total number of clients for firm A_i and N, the total number of consumers on the market.

Lemma 1. *Consumers are distributed between the three service providers according to the following rule:* $n_0 = N\left[p_1 + \beta q_1\right]$ *do not subscribe to any service,*
$n_1 = N\left[(p_2 - p_1) + \beta(q_2 - q_1)\right]$ *choose firm* A_1, $n_2 = N\left[(p_3 - p_2) + \beta(q_3 - q_2)\right]$
prefer firm A_2, $n_3 = N\left[1 - p_3 - \beta q_3\right]$ *select firm* A_3.

Proof of Lemma 1. We have $n_0 = c_1 N$, $n_1 = (c_2 - c_1)N$, $n_2 = (c_3 - c_2)N$ and $n_3 = (1 - c_3)N$. $\square$

3   Sharing Mechanism of the Grand Coalition'S Revenue and Stability

The game contains $2^3 - 1 = 7$ distinct coalitions. We assume that the game is with transferable utility i.e., the total revenue associated with each coalition can be freely shared between the coalition members. It implicitly means that the coalition members can freely give, receive or even, burn money [4]. Games with transferable utility require the introduction of a characteristic function $\nu : 2^3 - 1 \rightarrow \mathbb{R}$ which associates a global revenue with each coalition. Function ν valuates on a coalition, measures its worth.

To valuate the characteristic function on each coalition, we have chosen a representation by defensive equilibria [4].

Let $\mathcal{S}$ be the set of the 7 possible coalitions and $s \in \mathcal{S}$ a given coalition. We note $\sigma_s^{\text{opt}} = (p_i^{\text{opt}}, q_i^{\text{opt}})_{i \in s}$ where p_i^{opt}, $q_i^{\text{opt}} \in \mathbb{R}_+^2$, $\forall i \in s$ is the strategy on prices, QoS/content level chosen by the coalition s members and $\sigma_{\mathcal{S}-s}^{\text{opt}} = (p_j^{\text{opt}}, q_j^{\text{opt}})_{j \in \mathcal{S}-s}$ where p_j^{opt}, $q_j^{\text{opt}} \in \mathbb{R}_+^2$, $\forall j \in \mathcal{S} - s$ is the strategy selected by the firms who do not belong to coalition s. For each coalition $s \in \mathcal{S}$, we have to solve simultaneously $\sigma_s^{\mathrm{opt}} \in \arg \max_{\sigma_s = \left\{ (p_i, q_i) \in \mathbb{R}_+^2 \mid i \in s \right\}} \sum_{i \in s} U_i(\sigma_s, \sigma_{\mathcal{S}-s}^{\mathrm{opt}})$ and

$\sigma_{\mathcal{S}-s}^{\mathrm{opt}} \in \arg \max_{\sigma_{\mathcal{S}-s} = \left\{ (p_j, q_j) \in \mathbb{R}_+^2 \mid j \in \mathcal{S}-s \right\}} \sum_{j \in \mathcal{S}-s} U_j(\sigma_s^{\mathrm{opt}}, \sigma_{\mathcal{S}-s}).$

Using coalition s optimal strategy expression, we obtain the value of the characteristic function ν on coalition s: $\nu(s) = \sum_{i \in s} U_i(\sigma_s^{\mathrm{opt}}, \sigma_{\mathcal{S}-s}^{\mathrm{opt}})$ and on $\mathcal{S}-s$: $\nu(\mathcal{S}-s) = \sum_{j \in \mathcal{S}-s} U_j(\sigma_s^{\mathrm{opt}}, \sigma_{\mathcal{S}-s}^{\mathrm{opt}}).$

Theorem 1. *For each possible coalition, the two level game between the three providers admits a unique pure Nash equilibrium on prices and QoS/content level. The characterisitc function is then defined uniquely on each possible coalition.*

Proof of Theorem 1. For each coalition, we have to solve for each provider in the coalition, a two level game in price and QoS/content. Going backward, we start by determining the optimal prices for each provider in the coalition by differentiating the coalition's utility with respect to the prices. Then, we substitute the optimal prices in the coalition's utility and differentiate it once more, with respect to the QoS/content level. Finally, we obtain the optimal QoS/content level and the optimal prices for each provider in the coalition. □

In order to fix the ideas, we assume that $\mathcal{I}(.)$ is the identity application and that $\alpha_i = \vartheta(q_i) = \frac{q_i^2}{2}$. A cooperative game is an interactive decision model based on the behavior of groups of players or coalitions. In a cooperative game, one of the most difficult problem to solve is the sharing of the coalition's total revenue between its members. Shapley has proposed a fair sharing rule for a n player cooperative game. Another solution concept should require that the coalition members had no incentives to deviate to increase their revenue. Such a solution concept should guarantee the system *stability*. The set of the allocations of the grand coalition's revenue which should guarantee the system stability is the core of the cooperative game; formally, it is the set of the global revenue allocations $x = (x_i)_{i=1,2,\ldots,n}$ such that $\sum_{i=1,2,\ldots,n} x_i = \nu(1, 2, \ldots, n)$ (feasible) and $\sum_{i \in s} x_i \geq \nu(s), \forall s \subseteq \mathcal{S}$.

Computation of the Shapley value. The core of a cooperative game can be empty or very large. It explains partly, why this notion is so difficult to apply to predict the players' behavior. An alternative approach might be to identify a unique mapping $\phi : \mathcal{S} \to \mathbb{R}^3$ such that for the cooperative game defined by the characterisitc function ν, the expected revenue of each provider i is $\phi_i(\nu)$.

Theorem 2. *(Shapley [4]) There exists exactly one application $\phi : \mathcal{S} \to \mathbb{R}$ satisfying the four axioms of efficiency, symmetry, additivity and dummy player [4]. In a cooperative game with n players and transferable utility, this function satisfies the following equation for each player i and for any characteristic function ν: $\phi_i(\nu) = \sum_{\{s \subseteq \mathcal{S} \mid i \notin s\}} \frac{|s|!(n-|s|-1)!}{n!} \left(\nu(s \cup \{i\}) - \nu(s) \right).$*

Formula in Theorem 2 can be interpreted the following way. Suppose that we plan to assemble the grand coalition in a room but, the door to the room is only large enough for one player to enter at a time, so the players randomly line up in a queue at the door. The are $n!$ different ways that the players might be ordered in the queue. For any set s that does not contain player i, there are $|s|!(n-|s|-1)!$ different ways to order the players so that s is the set of players who are ahead of player i in the queue. Thus, if the various orderings are equally likely, $\frac{|s|!(n-|s|-1)!}{n!}$ is the probability that, when player i enters the room, he will find the coalition s there ahead of him. If i finds s ahead of him when he enters then, his marginal contribution to the worth of the coalition in the room when he enters is $\left(\nu(s \cup \{i\}) - \nu(s)\right)$. Thus, under this story of randomly ordered entry, the Shapley value of any player is his expected marginal contribution when he enters.

For each of the firms, we can compute the Shapley value which corresponds to a fair sharing of the grand coalition's total benefit between the firms. For firm A_i, we have $\phi_i(\nu) = \frac{1}{3}\left(\nu(A_i, A_j, A_k) - \nu(A_j, A_k)\right) + \frac{1}{6}\left(\nu(A_i, A_j) - \nu(A_j)\right) + \frac{1}{6}\left(\nu(A_i, A_k) - \nu(A_k)\right)$ where $i, j, k = 1, 2, 3$, $i \neq j$, $i \neq k$, $j \neq k$.

Influence of β on the system stability with Shapley value as sharing rule. We have represented the allocations given by the Shapley value for provider A_1, A_2, A_3, for different number of consumers on the market ($N = 5$, $N = 10$, $N = 50$, $N = 120$) as functions of $\beta \in [-1; 0]$. Each provider's allocation issued from the Shapley value are plotted on Figure 1 with information cost function being quadratic in the QoS/content level.

We observe that for small sensitivity coefficients, firms do not invest very much in QoS/content and prefer proposing small access prices. In this case, we observe in Figure 1, that A_3 dominates the market. Indeed, access prices being small, providers' revenues are small and hardly compensate the information cost which is quadratic in the QoS/content level. Thus, it is the favorite firm (A_3) who starts with a certain advantage and dominates the market.

For intermediate sensitivity coefficients, we observe in Figure 1, oscillations on the providers' revenues. Indeed, these latter are involved in a price war. Finally, when the sensitivity coefficient is high, providers invest massively in the QoS/content. The problem is then that the providers' access networks are limited in capacity. Thus, they cannot invest more than a fixed quantity $q_i^{\max}$, $i = 1, 2, 3$. For a large number of consumers, each provider invests at the maximum of his capacity and tries to differentiate with the access prices. At the end, the game converges when β decreases towards a perfect competition situation where the access prices are fixed such that the firms' profits are near 1 or even null.

Sufficient conditions guaranteeing the core existence. To determine the core of the game, we have to solve the following problem

$$(\mathcal{C}) = \Big\{(x_i)_{i=1,2,3} | \sum_{i=1,2,3} x_i = \nu(A_1, A_2, A_3), x_i \geq \nu(A_i) \; \forall i = 1, 2, 3, \; x_j + x_k \geq \nu(A_j, A_k) \; \forall j, k = 1, 2, 3, \; j \neq k \Big\}.$$

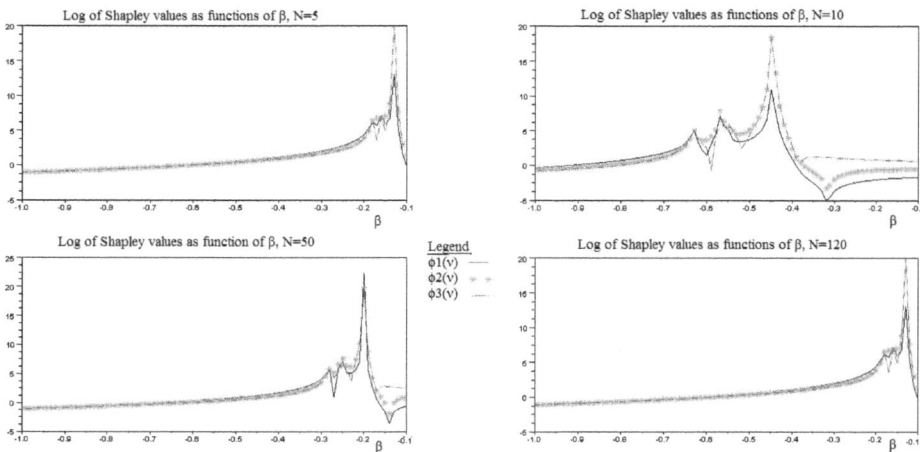

Fig. 1. Shapley value allocations for A_1, A_2, A_3 as functions of β with information quadratic in the QoS/content level

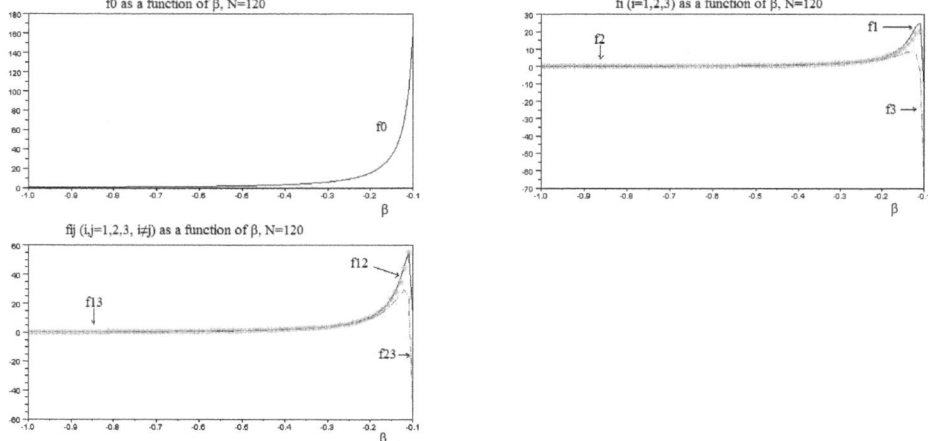

Fig. 2. Graphics of $f_0(\beta)$, $f_i(\beta)$, $f_{ij}(\beta)$, $i, j = 1, 2, 3$, $i \neq j$ as functions of β with information level being quadratic in the QoS/content level

Proposition 1. *If $\mathcal{I}(\alpha_i) = \frac{q_i^2}{2}$ and $\beta \in [-1; -0.78]$ then, if the three providers agree to share the grand coalition's total revenue according to the Shapley value, the grand coalition is stable i.e., no provider has incentives to leave it.*

Proof of Propositon 1. To start, we want to determine if there exists β values for which the Shapley value is in the core. To perform this point, we draw the 7 functions of β; each one of them describing System ($\mathcal{C}$) equality / inequalities. These functions are $f_0(\beta) = \phi_1(\nu) + \phi_2(\nu) + \phi_3(\nu) - \nu(A_1, A_2, A_3)$, $f_i(\beta)$

$= \phi_i(\nu) - \nu(A_i), \ i = 1, 2, 3, \ f_{ij}(\nu) = \phi_i(\nu) + \phi_j(\nu) - \nu(A_i, A_j), \ i, j = 1, 2, 3, \ i \neq j.$
Figure 2, we have plotted these 7 functions of β. We observe that the constraints defining the core of the game are satisfied if, and only if, $\beta \in [-1; -0.78]$ with an information level quadratic in the QoS/content investment level.   □

If the grand coalition is stable then, the regulatory authority's advices would not be followed by the providers. Indeed, the Shapley value being in the core of the game, the providers would rather follow this fair way to share the grand coalition's profit than deviating from it by following the regualtor's advices and risking to lose money. However, if the Shapley value is not in the core of the game, the providers' behaviors might be unpredictable and the regulator should intervene to control the system.

4   Conclusions

In this article, we have studied the game that arises between three service providers in competition, who have the opportunity to cooperate in order to increase their profits. We prove that if the consumers' sensitivity to the QoS/content coefficient (β) is inferior to -0.78 then, the Shapley value is a fair way to share the grand coalition revenue between the service providers and belongs to the core of the game i.e., provided the three service providers agree on this sharing rule the game remains stable. But if $\beta > -0.78$, the regulatory authority should intervene since the system behavior might be highly instable penalizing the consumers and the weakest providers. An extension should be to determine an upper bound on the interest rate of the repeated game such that under this threshold, consumers' welfare is guaranteed while market coverage is maximized.

References

[1] de Man, A.-P., Roijakkers, N., de Graauw, H.: Managing dynamics through robust alliance governance structures: The case of KLM and Northwest Airlines. European Management Journal (2010) (in press)

[2] Gulati, R., Khanna, T., Nohria, N.: Unilateral Commitments and the Importance of Process in Alliances. Sloan Management Review, MIT (1994)

[3] Le Cadre, H., Bouhtou, M.: An Interconnection Game between Mobile Network Operators: Hidden Information Forecasting using Expert Advice Fusion. Computer Networks Journal (2010) (in press),
http://dx.doi.org/10.1016/j.comnet.2010.05.007

[4] Myerson, R.: Game Theory, Analysis of Conflict, 6th edn. Harvard University Press, Cambridge (2004)

[5] Schlee, E.: The Value of Information About Product Quality. Rand Journal of Economics, The RAND Corporation 27, 803–815 (1996)

Business-Driven QoS Management of B2C Web Servers

Grażyna Suchacka[1] and Leszek Borzemski[2]

[1] Chair of Computer Science, Opole University of Technology
Sosnkowskiego 31, 45-272 Opole, Poland
g.suchacka@po.opole.pl
[2] Institute of Information Science, Wrocław University of Technology
Janiszewskiego 11/17, 50-370 Wrocław, Poland
leszek.borzemski@pwr.wroc.pl

Abstract. The paper deals with Quality of Service (QoS) in e-commerce Web servers. We consider request service control in a Business-to-Consumer (B2C) Web server system to ensure two business goals: high revenue of the e-retailer and high QoS for key customers with different business values. A problem of optimization of a multi-stage decision process in the context of achieving these goals is formulated. Some simulation results of the system performance under a novel admission control (AC) and scheduling algorithm are discussed.

Keywords: e-commerce, B2C, revenue, key customers, Quality of Service, QoS, Web server, scheduling, admission control.

1 Introduction

In the last two decades the Internet has experienced an explosive growth, which has been accompanied by a problem of low quality of service, especially with regard to WWW, which has often been referred to as the "World Wide Wait".

Due to the "best-effort" Internet service without request differentiation, it is not always possible to satisfy requirements of many up-to-date Web-based applications. For companies relying on the global network as business platform, service unreliability may have serious negative economic consequences. It is an incentive to develop QoS-enabled Web services by enhancing different network elements, especially Web servers, with the ability to provide different traffic classes with different QoS levels.

We focus on B2C Web servers, which allow individual users to browse and buy goods over the Internet. Guaranteeing proper QoS in such systems is the increasingly critical issue. Low QoS experienced by users strongly affects online retailers' revenues and has negative implications for companies trading online. Specificity of users' interaction with B2C Web sites, and particularly its financial aspect, indicate a necessity for considering Web server performance not only at the level of a computer system efficiency, but also from the perspective of business profitability.

The remainder of this paper is organized as follows. Section 2 overviews related work. Section 3 presents an idea of a novel approach to business-driven QoS management of a B2C Web server system, which model is described in Section 4. Section 5 provides formulation of the request service control problem in such a system.

B. Stiller, T. Hoßfeld, and G.D. Stamoulis (Eds.): ETM 2010, LNCS 6236, pp. 93–100, 2010.

Section 6 discusses some simulation results for the proposed approach and Section 7 concludes the paper.

2 Related Work

Although a plethora of work on AC and scheduling algorithms has been done so far, studies on B2C Web server systems have been limited. Some studies have addressed business-oriented QoS management of B2C Web servers. They have taken into consideration different business values of incoming requests, built in relation to different aspects of user sessions at a B2C Web site. A user session is defined as a sequence of temporarily and logically correlated requests issued by a user during a single visit to the site. It consists of some typical business operations, such as entry to the site, browsing and searching for goods, adding them to a shopping cart, etc. A crucial requirement for a B2C Web server is to support a user session integrity instead of relying on the service of individual HTTP requests, in order to enable users to continue their visits and to buy some goods.

Furthermore, there is a need for priority scheduling and admission control of requests with different business values. Most of approaches have proposed priorities based on a current session state, connected with a business operation performed by a user [1, 2, 3, 4, 5, 6]. Some approaches have used information on probabilities of transitions between the session states for different user profiles [2, 3, 5]. The next step towards business-oriented B2C service was taking into account financial values of goods in customers' shopping carts and observing a correlation between the session length and a probability of a product purchase [3]. On the other hand, a correlation of user's IP address with a fact of making a purchase in a given Web store has been used in an AC algorithm [7], motivated by the observation that returning customers have much bigger probability of making a purchase than users without prior purchases. A discount-charge model sensitive to user navigational structure and a corresponding AC and scheduling mechanism with load forecasting has been proposed in [8].

3 Novel Business-Driven QoS Management of a B2C Web Server

The literature analysis has shown that there is a need for preferential handling of the most valued customers of online stores, as in traditional retail trade. To the best of our knowledge, none of AC and scheduling algorithms for a B2C Web server have characterized the most valued customers nor have applied priority scheduling for them. We address this problem in combination with the problem of ensuring high revenue of the online retailer. With respect to these two goals, we propose an approach which is named KARO (*Key customers And Revenue Oriented admission and scheduling*), and the algorithm KARO-Rev (*KARO – Revenue Maximization*).

The idea of identifying the most loyal and profitable customers and their preferential handling in order to retain long-term relationships with them is the basis of many CRM (*Customer Relationship Management*) techniques. We propose applying this idea to the e-commerce environment and implementing it in a request service control mechanism for a B2C Web server system. We define key customers of the Web store as users who had made some purchases in the store at some time. We characterize them using RFM (*Recency, Frequency, Monetary*) analysis. It is a segmentation method which allows describing customers with RFM scores based on behavioral data related to past purchases: the time interval from the customer's last purchase until now, the total number of times the customer has made a purchase and the total amount of money spent by the customer. A way of performing RFM analysis for the use in our approach has been described in [9].

We propose creating a key customer database and storing it on the Web server. For each buyer there is a corresponding record in database, containing the properly coded behavioural data and the RFM score. We define two session classes: a *KC* class for key customers, and an *OC* class for ordinary customers. We propose describing all active sessions at the B2C Web site with session ranks. A *KC* session rank corresponds to the customer's RFM score read from the customer database, while an *OC* session rank is 0. Session ranks and other session attributes are used in KARO-Rev algorithm to determine control decisions on admission control and scheduling of requests belonging to different sessions.

4   Model of Request Service Control

We describe a request service control in a B2C Web server system as a multi-stage decision process. We consider the Web server system organized as a multi-tier application, consisting of a single Web server and a single back-end server (Fig. 1).

HTTP requests come to the Web server from the outside. Each Web page involves processing many HTTP requests, i.e. the first hit for an HTML page and following hits for all objects embedded in it. Requested Web objects may be static objects served by the Web server or dynamic objects created online.

Each incoming HTTP request i is classified in the context of belonging to a Web page p in a user session s and is denoted by x_{ip}^s. The request is also characterized by the request kind, $k_i \in \{0, 1\}$, where $k_i = 1$ for an HTML object, and $k_i = 0$ for an embedded object.

Accepted requests are placed in queue Q_1 in front of the Web server, where they wait for a service. If request's waiting time exceeds a predefined threshold T_1, the request will be dropped from the queue.

Dynamic requests are generated by the Web server and sent to the back-end server, if a response to the HTTP request has to be created online. A dynamic request is indexed with the dynamic request number j, the session number s, the page in session number p, and is denoted by y_{jp}^s. Dynamic requests are placed in queue Q_2 in front of the back-end server. Maximum request waiting time, after which a dynamic request will be dropped from the queue, is specified by a predefined threshold T_2.

Def. 1. Session s is considered to be *successfully completed* at the m-th moment, if an HTTP request x_{ip}^s completed at the m-th moment has been the last request of session s in the system and the response time for page p has not exceeded user page latency limit T_u.

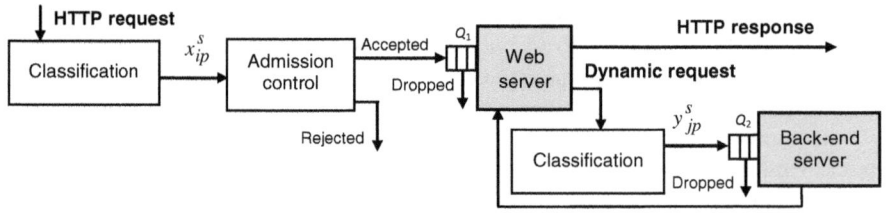

Fig. 1. The proposed model of request service in a B2C Web server system

Def. 2. Session s is considered to be *aborted* at the m-th moment, if a request belonging to session s has been rejected or dropped at this moment, or if after completing request x_{ip}^s at the m-th moment response time of page p exceeds T_u.

At any moment n there are $S(n)$ user sessions served by the system. We assume that each session $s = 1, 2, ..., S(n)$ is characterized by the following attributes:

- The session class $c^s(n) \in \{KC, OC\}$;
- The session rank $RFM^s(n) \in \{r\_min, ..., r\_max\}$;
- The session state $e^s(n) \in \{H, L, B, S, D, A, R, P\}$, corresponding to a type of business operation at the B2C site: entry to the <u>H</u>ome page (H), <u>L</u>ogging on (L), <u>B</u>rowsing (B), <u>S</u>earching for products (S), product's <u>D</u>etails (D), <u>A</u>dding a product to a shopping cart (A), <u>R</u>egistration (R) or making a <u>P</u>ayment (P);
- The session length $l^s(n) = 1, 2, ...$, meaning the number of pages visited in the session so far, including the current page;
- The financial value of products in a shopping cart $v^s(n)$, expressed in dollars;
- The state of a shopping cart $\eta^s(n) \in \{0, 1\}$ indicating the empty or not empty cart.

In order to support differentiated QoS in the Web server system, we propose introducing session priorities and applying a priority-based request AC and scheduling. Our session priority scheme is based on the method proposed in [3]. The session priority is updated at arrivals of HTTP requests for new Web pages only. We define two thresholds, T_{Med} and T_{Low}, which determine three ranges for the session length. The priority of session s at the n-th moment is calculated as follows:

$$P^s(n) = \begin{cases} Highest & \text{for } (c^s(n) = KC) \vee [(c^s(n) = OC) \wedge (\eta^s(n) = 1) \wedge (e^s(n) = P)], \\ High & \text{for } [(c^s(n) = OC) \wedge (\eta^s(n) = 1) \wedge (e^s(n) \# P)] \vee \\ & \quad [(c^s(n) = OC) \wedge (\eta^s(n) = 0) \wedge (l^s(n) < T_{Med})], \\ Medium & \text{for } (c^s(n) = OC) \wedge (\eta^s(n) = 0) \wedge (T_{Med} \leq l^s(n) < T_{Low}), \\ Low & \text{for } (c^s(n) = OC) \wedge (\eta^s(n) = 0) \wedge (l^s(n) \geq T_{Low}). \end{cases} \quad (1)$$

The priority *Highest* is reserved for all key customers, as well as for ordinary customers finalizing their purchase transactions. The priority *High* is assigned to all customers entering the site, to give a chance of successful interaction to all users, and to allow key customers to log into the site, as well as to ordinary customers with some items in their shopping carts. Priorities *Medium* and *Low* are assigned to ordinary customers who have stayed at the site for a long time with empty carts.

5 Key Customers Revenue Maximization Problem Formulation

We formulate a problem of optimisation of a multi-stage decision process, concerning the request service in a B2C Web server system in order to ensure high revenue and offer higher QoS to key customers with higher RFM scores. A Web server system is presented as an input-output control plant in an open loop system (Fig. 2).

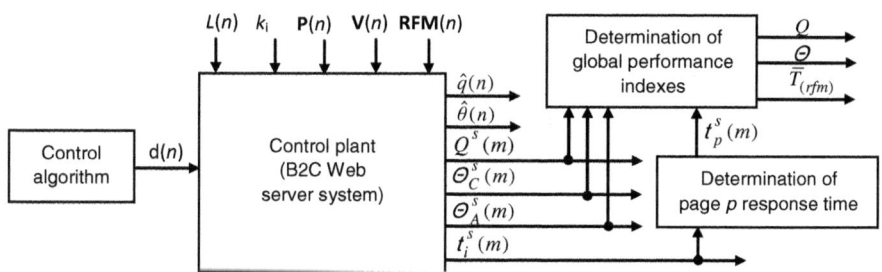

Fig. 2. B2C Web server system as a control plant

A control decision d(n), which is determined at arrival of an HTTP request x_{ip}^s or a dynamic request y_{jp}^s, may have one of three values: $a(x_{ip}^s(n)) = 0$ which means the rejection of the HTTP request; $z_1(x_{ip}^s(n))$ which means a position in queue Q_1 determined for the accepted HTTP request, and $z_2(y_{jp}^s(n))$ which means a position in queue Q_2 determined for the dynamic request.

Disturbances include current system load $L(n)$, the kind of HTTP request k_i, the vector of sessions' priorities $\mathbf{P}(n) = \left[P^1(n),...,P^{S(n)}(n)\right]^T$, the vector of shopping cart values $\mathbf{V}(n) = \left[v^1(n),...,v^{S(n)}(n)\right]^T$, and the vector of session ranks $\mathbf{RFM}(n) = \left[RFM^1(n),...,RFM^{S(n)}(n)\right]^T$ at the n-th moment.

Output variables may be divided into three groups:

I. Local performance indexes for the control decision d(n):

    1) $\hat{q}(n)$ - the potential revenue at the n-th moment:

$$\hat{q}(n) = \sum_{s=1}^{S(n)} v^s(n)\Big|_{\eta^s(n)=1},\qquad(2)$$

    2) $\hat{\theta}(n)$ - the vector of remaining request waiting times in queues Q_1 and Q_2.

II. Output variables computed after completion of HTTP request x_{ip}^s :

    1) $Q^s(m)$ – the amount of revenue achieved from session s at the m-th moment. It is equal to $v^s(m)$ if s is a session successfully completed at the m-th moment, for which $v^s(m) > 0$ and $e^s(m) = P$; in other cases it is equal to 0.

2) $\Theta_C^s(m)$ - the indicator of the successful completion of *KC* session *s*. It is equal to 1 if *s* is a *KC* session *successfully completed* at the *m*-th moment and 0 otherwise.

3) $\Theta_A^s(m)$ - the indicator of the failure of *KC* session *s*. It is equal to 1 if *s* is a *KC* session *aborted* at the *m*-th moment and 0 otherwise.

4) $t_i^s(m)$ - request response time for request *i* belonging to session s,

5) $t_p^s(m)$ - page response time for the *p*-th page in session s.

III. Global performance indexes for the observation window, in which *M* moments *m* occurred:

1) Q – the revenue throughput:

$$Q = \frac{\sum_{m=1}^{M} Q^s(m)}{number\_of\_time\_units} \tag{3}$$

2) Θ - the percentage of successfully completed *KC* sessions:

$$\Theta = \frac{100 * \sum_{m=1}^{M} \Theta_C^s(m)}{\sum_{m=1}^{M} \Theta_C^s(m) + \sum_{m=1}^{M} \Theta_A^s(m)} \tag{4}$$

3) $\overline{T}_{(rfm)} = [\bar{t}_{r\_min},...,\bar{t}_{(r\_max)}]$- the vector of 90-percentiles of page response times for different session ranks.

The optimal control decision $d^*(n)$ determined at the *n*-th moment maximizes the amount of potential revenue $\hat{q}(n)$ at this moment:

$$d^*(n) = \arg \max_{d(n)} \hat{q}(n) \tag{5}$$

subject to the following constraints:

$$(k_i = 1) \wedge [[(T_{AC_1} \le L(n) < T_{AC_2}) \wedge (P^s(n) = Low)] \vee [(L(n) \ge T_{AC_2}) \tag{6}$$
$$\wedge ((P^s(n) = Low) \vee (P^s(n) = Medium))]] \Rightarrow [d(n) = 0]$$

$$\forall a^{s_1}, b^{s_2} \in Q_k [(c^{s_1}(n) = KC) \wedge (c^{s_2}(n) = OC) \wedge (e^{s_2}(n) \ne P) \Rightarrow (a^{s_1} \prec b^{s_2})] \tag{7}$$

$$\forall a^{s_1}, b^{s_2} \in Q_k [(v^{s_1}(n) = v^{s_2}(n)) \wedge (RFM^{s_1}(n) > RFM^{s_2}(n)) \Rightarrow (a^{s_1} \prec b^{s_2})] \tag{8}$$

where T_{AC_1} and T_{AC_2} are predefined AC thresholds, a^{s_1} means request *a* belonging to sessions s_1 waiting in the queue Q_k, $k \in \{1, 2\}$.

The constraint (6) allows to limit the number of admitted HTTP requests under very high server load. The constraint (7) means that requests from *KC* sessions should be served before requests from *OC* sessions not connected with finalizing a purchase. The constraint (8) allows offering better QoS to more valued customers.

A sequence of control decisions $\{d^*(n)\}_{n=1}^{N}$ maximizes the revenue throughput in the observation window:

$$\{\mathbf{d}^*(n)\}_{n=1}^{N} = \arg\max_{\{\mathbf{d}(n)\}_{n=1}^{N}} Q,\tag{9}$$

as well as ensures the high percentage of successfully completed KC sessions Θ, and high differentiation between adjacent components of vector $\overline{T}_{(rfm)}$.

6 Simulation Results

Based on up-to-date results on real Web traffic characteristics, we worked out a simulation model of a B2C Web server system, including a session-based, HTTP-level workload model and a queuing network model of a multi-tiered Web server system [10]. We implemented the model in a discrete-event simulation tool using C++ and CSIM19 package. Using the simulator, an extensive performance study of the system performance under FIFO (*First-In-First-Out*) and KARO-Rev algorithm was performed. To compare relative benefits of our approach, exactly the same input workload was generated in the case of both policies. The results presented in this paper concern the following KARO-Rev parameter values: $T_1 = T_2 = T_u = 8$ seconds, $T_{Med} = 2$ pages, $T_{Low} = 20$ pages, $T_{AC_1} = 30$ and $T_{AC_2} = 80$ dynamic requests in the system. A share of KC sessions in the generated workload was 10% or 20%. Length of the observation window was 3 hours. A time unit was 1 minute.

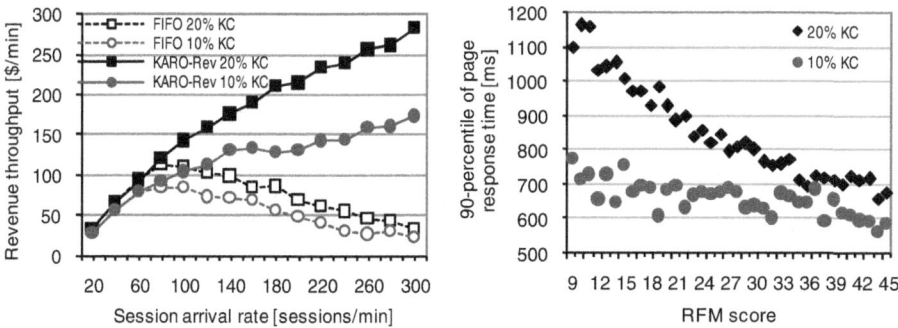

Fig. 3. Revenue throughput (left) and 90-percentile of KC page response time for different session ranks, provided by KARO-Rev for the session arrival rate of 300 sessions/min (right)

Fig. 3-left presents a variation of the revenue throughput as a function of the session arrival rate. Each point at the graf corresponds to a single experiment run for a constant session arrival rate per minute. The revenue throughput for KARO-Rev increases with the system load virtually throughout the whole load range, as opposed to FIFO, for which it starts decreasing at about 80 sessions generated per minute. Both for 10% and 20% of KC sessions, KARO-Rev has been also able to ensure very high percentage of successfully completed KC sessions, varying from 99.93 to 100.

Fig. 3-right presents the 90-percentile of KC page response time for different KC session ranks, ranging from 9 to 45, which has been provided by KARO-Rev for 300 sessions generated per minute. As it can be seen, the system tended to offer shorter page response times to sessions with higher ranks. Other results show that our algorithm has been also able to ensure significant improvements in overall system throughput and page response times for all sessions under overload.

7 Concluding Remarks

In this paper, a new approach to business-driven QoS management of a B2C Web server system is introduced and formally modeled. The proposed formal model may be the fist step to gain an analytical solution for the request service control problem in such a system. Simulation results show that the proposed KARO-Rev algorithm is able to guarantee business-oriented control goals, while considerably improving the overall system performance under overload. Future work include the improvement of the approach by considering QoS observed by a customer, including QoS latency. We also plan to use a prototype testbed to verify efficiency of KARO-Rev in more real conditions and to evaluate computational overhead incurred by the algorithm.

References

1. Carlström, J., Rom, R.: Application-aware Admission Control and Scheduling in Web Servers. In: Infocom 2002, vol. 2, pp. 506–515. IEEE Press, New York (2002)
2. Chen, H., Mohapatra, P.: Overload Control in QoS-aware Web Servers. Computer Networks 42(1), 119–133 (2003)
3. Menascé, D.A., Almeida, V.A.F., et al.: Business-Oriented Resource Management Policies for E-Commerce Servers. Performance Evaluation 42(2-3), 223–239 (2000)
4. Singhmar, N., Mathur, V., Apte, V., Manjunath, D.: A Combined LIFO-Priority Scheme for Overload Control of E-commerce Web Servers. In: IISW 2004 (2004)
5. Totok, A., Karamcheti, V.: Improving Performance of Internet Services Through Reward-Driven Request Prioritization. In: 14th IEEE IWQoS 2006, pp. 60–71. IEEE Press, Los Alamitos (2006)
6. Zhou, X., Wei, J., Xu, C.-Z.: Resource Allocation for Session-Based Two-Dimensional Service Differentiation on e-Commerce Servers. IEEE Transactions on Parallel and Distributed Systems 17(8), 838–850 (2006)
7. Yue, C., Wang, H.: Profit-aware Admission Control for Overload Protection in E-commerce Web Sites. In: 15th IEEE IWQoS 2007, pp. 188–193. IEEE Press, Los Alamitos (2007)
8. Shaaban, Y.A., Hillston, J.: Cost-based Admission Control for Internet Commerce QoS Enhancement. Electronic Commerce Research and Applications 8, 142–159 (2009)
9. Borzemski, L., Suchacka, G.: Discovering and Usage of Customer Knowledge in QoS Mechanism for B2C Web Server Systems. LNCS. Springer, Heidelberg (2010) (paper accepted)
10. Borzemski, L., Suchacka, G.: Web Traffic Modeling for E-commerce Web Server System. In: CCIS, vol. 39, pp. 151–159. Springer, Heidelberg (2009)

The Applicability of Context-Based Multicast - A Shopping Centre Scenario

Thomas Wozniak[1], Katarina Stanoevska-Slabeva[1],
Diogo Gomes[2], and Hans D. Schotten[3]

[1] Institute for Media and Communications Management, University of St. Gallen,
Blumenbergplatz 9, 9000 St. Gallen, Switzerland
{thomas.wozniak,katarina.stanoevska}@unisg.ch
[2] Instituto de Telecomunicações, Campus de Santiago, 3810 Aveiro, Portugal
dgomes@av.it.pt
[3] Chair for Wireless Communications and Navigation, University of Kaiserslautern,
67663 Kaiserslautern, Germany
schotten@eit.uni-kl.de

Abstract. This paper analyzes the applicability of context-based multicast content distribution (CBMCD) on the example of realistic push- and video-based mobile advertising services at a shopping centre. The technical results of the simulation of the service scenario show that CBMCD significantly reduces the number of unicast streams and the total volume of traffic in the network. The results of the financial analysis show that these technical benefits can be translated into considerable financial benefits due to costs savings. Taken together, these results suggest that CBMCD can be an efficient, cost-saving network traffic management approach and the basis for lucrative push services.

Keywords: Context Awareness, Multicast Content Delivery, IP Multicast.

1 Introduction

Global mobile data traffic doubles every year between 2009 and 2014 [1]. Mobile data traffic and revenues have become decoupled [2], which forces operators to make their networks more efficient and to reduce network costs. Multicast content delivery can improve the efficiency of existing networks. With multicast, one data stream is shared by a group of subscribers who receive the same content. This is contrary to unicast content delivery where one data stream is established for each subscriber. Thus, multicast solutions can considerably reduce the network load in scenarios where subscribers receive the same content. The reduction comes at the cost of more limited service, which needs to be shared amongst users. The challenge is therefore for the operator to detect and create content and services that can be shared by several users. One such way is through contextualization of users and the grouping of such users based on their shared interests. The identification of a sufficient number of subscribers who qualify to receive the same content and who thus form a multicast group becomes a prerequisite for making multicast content delivery viable. This prerequisite can be met by acquiring and processing subscribers' context. Context is any

B. Stiller, T. Hoßfeld, and G.D. Stamoulis (Eds.): ETM 2010, LNCS 6236, pp. 101–108, 2010.
© Springer-Verlag Berlin Heidelberg 2010

information that can be used to characterize the situation of a subscriber that is considered relevant for delivering multicast content [3]. Examples of such context information are situation-specific information (e.g. location, weather, and social relationships) and user profile information (e.g. gender, age, and interests).

Besides the multicast service requirements, there are transmission requirements. Multicast transmission of data streams, especially in the wireless environment, implies several technology tradeoffs. One of the most important tradeoffs is the power control mechanism of the Radio Interface that might limit the coverage of a radio cell and/or throughput if a multicast transmission is established. Therefore wireless technologies have power control algorithms that control the tradeoff between reusing the same channel to deliver data and the cost associated with that. In technologies such as 3GPP MBMS, this factor has been thoroughly studied in [4], [5], and [6]. These studies point towards the need for at least 5 terminals to share the same data stream in order for multicast to be technically advantageous. Furthermore, network equipment needs to support multicast functionalities, which usually do not come as standard part of infrastructure and result in higher costs. Because multicast content delivery costs around three to five times as much as unicast, multicast can only improve network efficiency and be financially profitable if it serves a group of subscribers of a certain size. The research project C-CAST developed a solution that combines group formation based on user context with multicast content distribution, enabling Context-Based Multicast Content Distribution (CBMCD) [7].

This paper analyzes the applicability of CBMCD on the example of a Swiss shopping centre. The main contribution of this paper is twofold: on the one hand, it quantifies potential benefits of CBMCD on the example of a real-world scenario; on the other, it translates these technical results into potential financial benefits. The remainder of this paper is structured as follows. Section 2 describes the research methodology. Section 3 describes the simulated use case scenario. Section 4 presents the results of the simulation and the financial impact calculation. Section 5 concludes with a summary of results, limitations of research, and further research directions.

2 Methodology

This paper follows a two-step approach. First, a use case scenario for CBMCD in a shopping centre is simulated. The simulation shows how real-world service subscribers would behave in such a scenario and how the network would be used to provision the services. Second, potential financial benefits due to CBMCD are calculated. The simulation results serve as input for the financial impact calculation.

The used simulator is an advanced Geographic Information System (GIS) based on the simulator described in [8]. It takes three vectors as input: network properties, terminals, and services distribution and properties. The first vector includes the location and the capabilities of each base station including coverage and bandwidth capabilities. The second vector contains the location of the terminals. The last vector contains the available services, their average penetration ratios (in relation to the overall population of terminals), and the services' traffic properties. The simulator relates the three vectors by creating a time event in which the terminals are consuming the services in the defined network and calculates the associations of terminals to the base stations by considering the coverage characteristics of the base stations and decisions of the radio resource management. Based on the service penetration ratios, the simulator randomly associates services to terminals. Based on this association, the network load is calculated for each of the base stations, providing a snapshot of the network load. The simulator output includes the service distribution across terminals (calculation based on the average penetration ratios) and the distribution of services per base station (number of terminals and bandwidth used).

Simulation results provide both legacy (unicast-based transmission) and CBMCD views of how services are delivered. CBMCD results are calculated according to what is most technically efficient from the viewpoint of power-control mechanisms in the cell. In this paper, the considered switching factor between unicast and multicast is 5 in accordance to the studies in [4], [5], and [6].

The simulation results allow the calculation of the amount of traffic transferred in the scenario through unicast and multicast. The number of subscribers served via multicast is input to the evaluation of the financial impact of context-based multicast services in the scenario.

3 The Use Case Scenario of the Shopping Arena St. Gallen

Mobile advertising at the Shopping Arena St. Gallen (SASG) was chosen as a use case. With an area of around 32,000sqm for shops and restaurants, the SASG is the largest shopping centre in Eastern Switzerland [9], [10]. In 2009, the SASG had 3.7 million visitors and generated a turnover of around €145 million [10].

The decision to choose mobile advertising in a shopping centre as a use case scenario was motivated by the optimistic predictions for the mobile advertising market (e.g. [11]) and the high suitability of shopping centres for both retail-oriented mobile advertising and location-specific CBMCD. A further reason is the high potential of mobile advertising for all involved players: the shopping centre and its shops, which can access customers in an innovative and personalized way; the customers, who get informed about special offers and coupons; and telco operators, for which it can become a lucrative service.

Shopping centres are well suited for mobile advertising for three reasons. First, shopping centres typically bring together a large number of shops, which wish to advertise in order to draw customers. Second, because people visit shopping centres primarily to do shopping, they are likely to be relatively more inclined to receive mobile ads. Third, shopping centres have per se a high probability to be suitable for multicast, as they attract a lot of people at the same time with high probability that many of them have similar shopping interests.

The values for the three input vectors required for the simulation tool are based on realistic data about available network infrastructure, number and temporal distribution of SASG visitors, and opt-in behaviour for mobile advertising campaigns.

3.1 Network and Terminals Properties

Six cellular base stations were identified nearby the SASG [12]. Because we propose a future service, we assume that all six base stations feature UMTS. The number of terminals is estimated by drawing upon SASG visitor statistics. Considering 3.7 million visitors and 307 shopping days in 2009 [10], the SASG had an average of

Table 1. SASG visitor frequency by type and time of day

Type of day	Time periods (length)	Average number of visitors per hour
Weekdays	9am - 5pm (8 hours)	904 [a]
	5pm - 7pm (2 hours)	1,808 [b]
Saturdays	9am - 1pm (4 hours)	1,808 [c]
	1pm - 5pm (8 hours)	2,712 [d]

12,052 visitors per day or 72,313 visitors per week, respectively. The varying visitor frequency by day and by time of day is considered by distinguishing two types of days (weekdays and Saturdays) and a period of low and one of high visitor frequency for each type of day. The specific time periods were chosen based on SASG opening hours [9] and SASG visitor statistics [10]. The average number of visitors per week was distributed over the defined time periods assuming that the visitor frequencies in these periods are related to each other as follows: $b = 2 * a$, $b = c$, and $d = 1.5 * c$. The results are shown in Table 1. Each visitor is assumed to have a 3G-capable mobile terminal.[1] The terminals were distributed around the geometric centre of the SASG using a Normal distribution.

3.2 Services Distribution and Properties

The services proposed for this scenario are push- and video-based mobile advertising services defined as a number of repeated ad messages for a specific category of products or services. Given that video is among the most preferred mobile advertising media formats [14], video ads of 60 seconds in length that are streamed at 256kbps to a specific target group every hour were chosen as services. Seven shop categories currently present in the SASG were considered as customers: fashion and shoes, groceries, furniture, books, electronics, health, and restaurants [9]. As a result, there are seven different services, each of which serving video ads for products or services in one of the aforementioned categories. To determine the distribution of the services in the total population of SASG visitors, we correspond each service to a specific group of visitors, contingent on gender and age group (grouping of users based on their context information).

Major obstacles for the acceptance of mobile advertising are fear of unsolicited ad messages and lack of perceived control [15]. Thus, a SASG visitor needs to opt in to receive mobile ads that are relevant to him or her. We assume that, on average, 25% of all SASG visitors opt in. This is based on previous studies on mobile advertising [16], [17]. Further, we assume that the portion of visitors who opt in decreases by age. Message frequency is among the most important opt-in contract requirements [18]. We limit the number of received ad messages to one per hour.

Table 2 shows the seven services, gender and age of the corresponding groups of visitors, the size of these groups relative to the total population[2], and the numbers of visitors who are assumed to opt in to receive mobile ads.

---

[1] In Q8 2008, mobile penetration in Switzerland was 117.1% [13]. Since we propose a future service, we assume all mobile subscribers have moved to 3G.

[2] The gender and age distribution of the SASG visitors is assumed to correspond to that of the overall Swiss population [19].

Table 2. Services Distribution

Service	Gender and age of corresponding visitor group	% of total pop.[4]	No. of visitors who opted in		
			904	1808	2712
Fashion and shoes	Females, 20+	40.6%	77	154	232
Groceries	All	100.0%	221	442	663
Furniture	Males, females, 20+	79.0%	154	309	463
Books	Males, females, 20+	79.0%	154	309	463
Electronics	Males, 20+	38.4%	77	154	232
Health	Males, females, 40+	52.4%	70	140	210
Restaurants	All	100.0%	221	442	663

4 Discussion of Results

The purpose of the simulation was to study the impact of using Multicast to distribute content to subscribers based on their context for the scenario described in the previous section. The simulation results provide scenario-based service level information that is further processed in this section into a technical and financial analysis.

4.1 Technical Analysis

Three scenarios were studied corresponding to average visitor rates in a low, medium, and high influx period. Fig. 1 illustrates that far less streams are required in a context-based unicast/multicast service delivery scenario as proposed in this paper, compared to the current day scenario where streams are only delivered through unicast streams. From the same figure, it is also clear that as the average number of visitors increases, the number of context-based unicast streams decreases and the number of multicast streams increases. The initially higher value of unicast streams is explained by low penetration of some of the cells. In such cells, there are initially no subscribers. As visitors flux increases, services penetrate all the cells and ultimately all visitors will receive the service through multicast. In an extreme scenario, the number of legacy unicast streams is equal to the number of subscribers. In the proposed solution, the scenario will evolve into a full multicast scenario in which the number of streams is

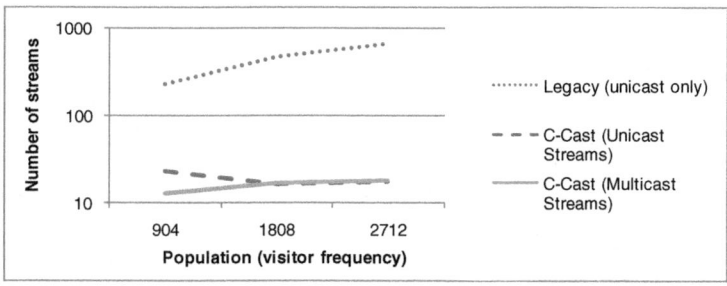

Fig. 1. Number of streams dependent of visitor frequency

the number of services times the number of cells (7 * 4 = 28). In such a densely popu-
lated area, a legacy unicast service deployment will inevitably lead to a very high
blockage probability and consequent loss of revenue due to the inability to serve
the service.

From the simulation, we can also extrapolate the number of bits transferred. We
consider that all services consist of a video file of 60 seconds in length streamed at
256kbps. In a legacy scenario (unicast only), the number of transferred bits grows
linearly with increasing number of terminals. That growth is insignificant in terms of
Unicast plus Multicast streams associated to the CBMCD. In comparison, CBMCD
reduces the number of bits transferred by 83.93%, 92.93%, and 94.71% for the low,
medium, and high visitor flux periods, respectively. Another interesting result from
the simulation is the average group size per cell, which grows linearly from 14 to 34.
This result is relevant from a technical viewpoint, since it shows that the switching
factor of 5 is quickly overcome.

4.2 Financial Analysis

The previous section has shown that CBMCD significantly reduces the number of
unicast streams and the total volume of transmitted bits. This section describes the
financial impact of such bandwidth savings.

The videos ads are paid by the advertising shops per each received video. For simplicity, the price charged for each received video is the cost price for a unicast video.
Based on [24], the price per video can be calculated to be 11.25 Euro cents.[3] Because
all unicast videos are sold at cost price, no financial benefit can be derived from them.
Thus, unicast videos are not considered any further. The financial benefit that can be
derived from multicast videos depends on three factors: the costs of streaming a multicast video, the number of multicast streams, and the number of visitors who receive
a multicast stream. Multicast streaming is around four times as expensive as unicast
streaming. Thus, one multicast video costs 45 Euro cents. The number of multicast
streams and the number of multicast stream receivers are shown in Table 3. Table 3
also shows the revenues that can be derived from the multicast videos, the costs incurred by streaming the multicast videos, and the resulting financial benefits.

Table 3. Results of financial analysis

Population	904	1808	2712
Number of Multicast Streams	13	17	18
Number of Multicast video receivers	201	451	645
Revenues from multicast videos	€ 22.61	€ 50.74	€ 72.56
Costs of multicast videos	€ 5.85	€ 7.65	€ 8.10
Benefit from multicast per hour	€ 16.76	€ 43.09	€ 64.46

The financial benefits shown in Table 3 are only snapshots reflecting one hour in a
period with a specific visitor frequency. These snapshots are extrapolated to obtain
the financial benefit for one week, month, and year. The financial benefit for one
week is calculated by considering length (in hours) and frequencies (five weekdays

---

[3] 11.25 Euro cents = 60sec video stream * 256kBits/sec * 0.000000715256 Euro cents/bit [2].

and one Saturday) of the snapshots (see Table 1). The resulting financial benefits per week, month (week times four), and year (month times twelve) amount to € 7'650, € 30'600, and € 367'195, respectively. In one year, the financial benefit of multicast in the given scenario is € 367'195. This amount contributes to offsetting fixed costs (e.g. for marketing staff and the initial investment required for the multicast setup).

5 Conclusion

This paper has analyzed the applicability of CBMCD on the example of a Swiss shopping centre. Push- and video-based mobile advertising services at this shopping centre were chosen as a use case scenario. The technical results of the simulation of the use case scenario show that CBMCD significantly reduces the number of unicast streams and the total volume of traffic in the network. The results of the financial analysis show that these technical benefits can be translated into considerable financial benefits. Taken together, these results have two main implications: first, locations with similar characteristics as shopping centres can be suitable for lucrative push-CBMCD services; and second, telco operators have reason to consider CBMCD and mobile advertising for tapping new revenues sources.

Even though the use case of the shopping centre in St. Gallen provided a realistic setting for the research, there are limitations that need to be considered and that require further research in order to generate more generalizable results. For example, from the perspective of a telco operator, it might be of interest to have indicators for choosing suitable locations where to offer push-CBMCD services. Such indicators could be the minimum user frequency required at the specific site or the required frequency of push services to achieve savings. Furthermore, the calculation of financial benefits only considered potential revenues and costs of transferred bits. However, CBMCD requires investments and adjustment of the existing infrastructure. Given these limitations, further research is required to investigate potential minimum and maximum values of the considered parameters. Generalized findings can be included in automatic decision-supporting tools for better choice between unicast and multicast in practice.

Acknowledgments. The research presented in this paper was carried out in the C-CAST project, which is supported by the European Commission under the grant no. ICT-2007-216462.

References

1. Cisco: Cisco Visual Networking Index: Global Mobile Data Traffic Forecast Update, 2009-2014. Whitepaper (February 9, 2010), http://www.cisco.com/
2. UMTS Forum: Mobile Broadband Evolution: the roadmap from HSPA to LTE. Whitepaper (February 2009),
   http://www.umts-forum.org/content/view/2693/174/
3. Dey, A.K., Abowd, G.D.: Towards a better understanding of context and context-awareness. GVU Technical Report GIT-GVU-99-22, College of Computing, Georgia Institute of Technology, ftp://ftp.cc.gatech.edu/pub/gvu/tr/1999/99-22.pdf

4. Alexiou, et al.: Power Control Scheme for Efficient Radio Bearer Selection in MBMS. In: IEEE International Symposium on a World of Wireless, Mobile and Multimedia Networks, WoWMoM 2007, pp. 1–8 (2007)
5. De, V., et al.: Multimedia broadcast and multicast services in 3G mobile networks. Alcatel Telecommunications Review (2003)
6. IST-2003-507607 (B-BONE), Deliverable of the project (D2.5), Final Results with combined enhancements of the Air Interface, http://b-bone.ptinovacao.pt
7. C-CAST, ICT-2007-216462, http://www.ict-ccast.eu/
8. Gomes, D.: Optimização de recursos para difusão em redes de próxima geração. Phd Thesis, Univ. Aveiro, Portugal (2010)
9. Shopping Arena St. Gallen, http://www.shopping-arena.ch/
10. Shopping Arena St. Gallen, Shopping Arena auf Kurs, Media release (January 7, 2010), http://www.shopping-arena.ch/documents/ MM_Rueckblick_070110.pdf
11. Juniper Research: Advertising ~ The future's bright, the future's mobile. Whitepaper (June 2009), http://juniperresearch.com/
12. Swiss Federal Office of Communications, Location of radio transmitters in Switzerland, http://www.funksender.ch/webgis/bakom.php
13. Swiss Federal Communications Commission ComCom, Mobile telephony, http://www.comcom.admin.ch/dokumentation/00439/00467/ index.html?lang=en
14. Barnes, S.J., Scornavacca, E.: Uncovering patterns in mobile advertising opt-in behaviour: a decision hierarchy approach. International Journal of Mobile Communications 6(4), 405–416 (2008)
15. Dickinger, A., Kleijnen, M.: Coupons going wireless: Determinants of consumer intentions to redeem mobile coupons. Journal of Interactive Marketing 22(3), 23–39 (2008)
16. Barwise, P., Strong, C.: Permission-based mobile advertising. Journal of Interactive Marketing 16(1), 14–24 (2002)
17. Nielsen Mobile: Mobile Media Europe: State of the EU5 Union. Research Showcase on Mobile Advertising, IAB Belgium, Brussels (March 16, 2009), http://www.iabeurope.eu/
18. Bamba, F., Barnes, S.J.: SMS advertising, permission and the consumer: a study. Business Process Management Journal 13(6), 815–829 (2007)
19. Swiss Federal Statistical Office: Statistik des jährlichen Bevölkerungsstandes (ESPOP) und der natürlichen Bevölkerungsbewegung (BEVNAT) 2009 - Provisorische Ergebnisse (February 25, 2010), http://www.bfs.admin.ch/

Author Index

GPSR Compliance

The European Union's (EU) General Product Safety Regulation (GPSR)
is a set of rules that requires consumer products to be safe and our
obligations to ensure this.

If you have any concerns about our products, you can contact us on
ProductSafety@springernature.com

In case Publisher is established outside the EU, the EU authorized
representative is:

Springer Nature Customer Service Center GmbH
Europaplatz 3
69115 Heidelberg, Germany

Batch number: 09474016

Printed by Printforce, the Netherlands